The New
Wider World
Second Edition

TEACHER'S RESOURCE BOOK

Neil Punnett and Alison Rae

Nelson Thornes

a Wolters Kluwer business

Text © Neil Punnett, Alison Rae 2003
Original illustrations © Nelson Thornes Ltd 2003

The rights of Neil Punnett and Alison Rae to be identified as authors of this work have been asserted by them in accordance with the Copyright, Designs and Patents Act 1988.

All rights reserved. The copyright holders authorise ONLY users of *The New Wider World Second Edition Teacher's Resource Book* to make photocopies of the maps and activity sheets for their own or their pupils' immediate use within the teaching context. No other rights are granted without permission in writing from the publishers or under licence from the Copyright Licensing Agency Limited. Further details of such licences (for reprographic reproduction) may be obtained from the Copyright Licensing Agency Limited, of Saffron House, 6–10 Kirby Street, London EC1N 8TS.

Copy by any other means or for any other purpose is strictly prohibited without prior written consent from the copyright holders. Application for such permission should be addressed to the publishers.

Any person who commits any unauthorised act in relation to this publication may be liable to criminal prosecution and civil claims for damages.

Published in 2003 by:
Nelson Thornes Ltd
Delta Place
27 Bath Road
CHELTENHAM
GL53 7TH
United Kingdom

07 08 09 10 / 10 9 8 7 6 5 4

A catalogue record for this book is available from the British Library

ISBN 978 0 7487 7377 0

Page make-up and illustrations by IFA Design, Plymouth

Printed in Croatia by Zrinski

Acknowledgements
With thanks to the following for permission to reproduce photographs and illustrations (referred to by page number) and other copyright material in this book:

Janet Elkes: pages 200–201; **Flying Flowers:** page 80; **Hutchison:** page 63; **Geoff Mackley/Reuters:** page 152.

Oxford University Press for text extract from *The Third World*, R. Beddis, on page 25; *Department of Statistics, Malaysia* for statistics on page 28; **Larousse** for statistics from *Francoscopie 2001*, G. Mermet, on pages 30–32; *The Guardian* for extract from 14 July 1995 on page 52; **Understanding Global Issues Ltd** for the extract on page 76; *The Independent* for the extract from 5 February 1997 on page 79; *Peeblesshire News* for the extract from 26 September 2002 on page 92; **www.responsibletravel.com** for the extract on page 121; **Ing K** for the extract on page 122; 'When the tourists flew in' by Cecil Rajendra on page 122; **UNICEF** for extracts from the reports *The State of the World's Children 1995* and *2000* on pages 125–126; **Routledge** for the extract from *The Atlas of African Affairs (Second edition)* by I. Griffiths, on page 129; *Glasgow Herald* for the extract from 2 January 2003 on page 131; **The Meteorological Office** for the extract on page 139; *The Independent* for the extract from 8 June 1992 on page 150; *The Glasgow Herald* for the extract from 2 January 2003 on page 151; **http://ccs.cla.kobe-u.ac.jp/Asia/Visitor/Furm/** for the extracts on pages 179, 184–185; **www.olywa.net/radu/valerie/mshafter.html** for the extract on page 183; *USA Today* for the extract from 11 May 2002 on page 183; **www.rediff.com** for the extract on pages 187–188; *The Observer* for the extract from 1 December 1996 on page 211.

Every effort has been made to contact copyright holders. The publishers apologise to anyone whose rights have been inadvertently overlooked and will be happy to rectify any errors or omissions.

Contents

Teacher's Notes

Introduction

This new edition of *The New Wider World Teacher's Resource Book* is intended to complement and enhance the material in *The New Wider World, Second Edition,* helping the busy Geography teacher to save time by providing instant, accessible ideas and activities and to extend their pupils' learning through opportunities for more in-depth study and investigation of topics and issues. Many of the activity sheets can stand alone, but they work best as complements to the pupil book. They can be used within a planned scheme of work or as revision exercises.

Structure of the *Teacher's Resource Book*

This new *Teacher's Resource Book* has been written to provide a wide range of new resources for teachers to use during the course. The 19 chapters of the book are linked directly to the new pupil book; the 📖 symbol on the Chapter Commentary pages indicates the relevant pages in *The New Wider World, Second Edition*. Each chapter adopts the following approach.

Chapter Commentary

Each chapter commentary summarises the content of the chapter in the pupil book and includes some ideas on how to extend the work presented there. The resource activity sheets in the *Teacher's Resource Book* are linked to the relevant pages in the pupil book with ideas for making the most of the activity sheets to help support the teacher. Some of the activities are more demanding than others and these are highlighted in the chapter commentary. The extra case study is also linked to the case study in the pupil book and its complementary or, more often, contrasting nature is elucidated in the chapter commentary.

Activity Sheets

Each chapter has five activity sheets (except Chapter 6, with three) containing a variety of questions and activities linked to specific sections within the pupil book. There are a number of activities involving the use of ICT. The activity sheets feature a variety of stimulus materials including bar graphs, line graphs, pie graphs, scattergraphs, sketches, maps, newspaper articles, raw data, cartoons and photographs.

Case Study Extra

This extra case study material has been carefully chosen to build on the case study in the pupil book and to offer ideas for alternative examples and investigations – usually it offers a contrasting example.

Exam Practice Questions

Each Case Study Extra concludes with GCSE-style questions to give opportunities for practice in answering questions on case study material in an examination style.

Improve Your Mark!

Model answers are provided for selected exam practice questions in each Case Study Extra with additional notes on how pupils can improve their answers and increase their marks.

An example of the way in which the activity sheets relate to the pupil book is provided by Activity Sheet 7.4 on dairy farming in Somerset. The pupil book describes the factors that affect the choice of farming type and goes on to look briefly at the distribution of the main farming types in the UK. The activity sheet expands on this by studying one example in greater depth – dairy farming. Starting with a map of the main dairy farming areas in the UK, the focus of the activity sheet is to explore the influence of climate on dairy farming in Somerset. A table showing monthly average temperature and rainfall provides the basis for three questions including the completion of a climate graph. Finally a question is posed on the influence of the EU's common agricultural policy on UK dairy farming.

A second example of the way in which the activity sheets relate to the pupil book is provided by Activity Sheet 16.3 on the Kobe earthquake. The activity sheet opens with questions based directly on the case study on pages 270 and 271 of the pupil book. The book provides excellent diagrams to explain the cause of the earthquake and goes on to list its effects, with two impressive photographs, and what happened in the months after the earthquake. The main aim of the activity sheet is to place some flesh on the bone of the effects of the earthquake and the approach taken is to provide an actual eyewitness account of the earthquake written by a Japanese university student. The immediacy and drama of this description brings the pupil book's list of effects to life. The chapter commentary notes for Activity Sheet 16.3 in this *Teacher's Resource Book* provides a good example of the additional assistance the commentaries provide for the busy teacher using the activity sheets.

In summary, *The New Wider World Teacher's Resource Book* is intended to assist Geography teachers in their work and to extend their use of the materials in *The New Wider World, Second Edition* by helping their pupils to gain the skills, knowledge and understanding required for their GCSE Geography examination.

Assessment

Each chapter in this *Teacher's Resource Book* contains a wide variety of forms of assessment, in both the activity sheets and the exam practice questions.

The theme of each activity sheet is closely linked to those in the relevant chapter of the pupil book and the materials provided on each sheet are designed to complement those in the pupil book whilst illustrating the same or related points and issues. This gives pupils extra practice as well as highlighting key points. In some instances activity sheet questions utilise diagrams or data in the corresponding chapter of the pupil book. A good example of this is Question 2 on Activity Sheet 1.1, which is based on resource material from page 4 of the pupil book. Similarly, Activity Sheet 5.3 bases its Figure A on Figure 5.10 in the pupil book.

Supporting GCSE requirements

The *Teacher's Resource Book* is designed to help candidates with revision and other forms of exam preparation in a number of ways.

By providing a large resource of extra practice questions suitable for both lesson and homework time

Many questions in the *Teacher's Resource Book* are phrased so that they resemble typical GCSE-style questions as closely as possible. Obviously, the more practice pupils have using these, the better. Familiarity with the main command words is essential for accurate response to questions. For example, Activity Sheet 1.1 includes four key command words – define, list, identify, annotate – within six questions, as well as requesting 'advantages and disadvantages'. In Activity Sheet 5.3 pupils must learn to distinguish between 'compare' and 'contrast' in Question 2 and between 'describe' and 'account for' in Question 3.

Activity sheets do vary in length, giving scope to choose something suitable for a particular time period available in class or during homework.

By providing plenty of extra material, both in the activity sheets and the Case Study Extras

By using and exploring case studies in depth pupils eventually have these at their fingertips and are more able to apply them to exam questions. Knowledge of specific case studies is needed especially in longer 'levels' questions, where because the mark allocation can be as high as nine or ten lots of detail must be included. The Case Study Extra material has been selected so that it complements the case studies in *The New Wider World, Second Edition*. For example, in Chapter 14 of the pupil book the case study ecosystems concerns sustainable forestry in Malaysia; the *Teacher's Resource Book* explores the sand dune system at Shell Bay in Dorset. Therefore two different ecosystems in diverse parts of the world, in countries in different states of economic development, have been tackled between the two books.

Various activities also reinforce and extend the case study materials. For example, Activity Sheet 2.4 includes an example of out-migration from a Turkish village; this supports the case study on Turks migrating into Germany on pages 28 and 29 of the pupil book.

By approaching the material from another angle, so testing whether pupils have completely understood the issues and arguments

Other questions in the *Teacher's Resource Book* are designed to increase pupils' familiarity with the course material. By looking at familiar material in a new way, understanding is increased. One example of this can be found on Activity Sheet 2.3, which deals with 'push' and 'pull' factors. The same situation is looked at from the points of view of two very different households, so the pupils see that people may not all react in the same way to a given situation. Exam Practice Question 2 in Chapter 7 of the *Teacher's Resource Book* requires pupils to transfer what has been done in the Case Study Extra to another market gardening case study they know.

By employing a broad variety of types of data, so that pupils will not be unnerved by any approach an exam question takes

A very broad range of data sources has been utilised in the preparation of this *Teacher's Resource Book*. To give a few examples, Activity Sheet 11.1 includes UNICEF data; 8.5 refers to a local newspaper, the *Peeblesshire News*; the Case Study Extra in Chapter 13 uses the *Glasgow Herald*; and Activity Sheet 14.1 uses *Geofile* 431 as a source.

By practising a wide range of skills

Some activity sheets require pupils to use other resources as extra data sources. For example, Question 2b(i) on Activity Sheet 11.5 on Aid asks for use of an atlas. Sometimes figures in the pupil's book are used for reference. While this approach adds a breadth to the material used by the sheets, it also trains pupils to handle several different data sources at once, in a similar way to some of the longest 'levels' questions in GCSE examinations.

A few activity sheets employ photographs. Activity Sheet 7.3 links a picture of an agricultural landscape in France to the technique of field sketching, supporting fieldwork and coursework skills. This sheet also involves pupils in constructing a systems diagram for the farm in the photograph. Various maps are also included – sketch maps as in Activity Sheet 3.1 and Ordnance Survey maps as in Activity Sheet 3.2. Scaled symbols, such as flow lines, appear in a number of activities such as Activity Sheet 3.4 and there are several types of graph throughout the chapters.

By training pupils to evaluate their own performance constantly and so improve the quality of their work automatically

Each Improve Your Mark! section includes detailed advice to pupils in how to improve their own performance. These sections deal primarily with the 'levels'-type questions, which are so important for pupils to master to be able to achieve top grades.

At least one model answer is provided, coupled with comments and close guidance on how to write at Level 3. The increase in detail and quality of answers from Level 1 through to Level 3 is set out. Pupils are invited to compare their answers to this material and assess their level. Then they can improve their answers to bring them up to the top level. After tackling a number of questions like this pupils will begin to review their work automatically as they write and so will take this new skill into the exam room with them!

Moreover, this review and evaluation skill will also benefit coursework. Evaluation counts for a proportion of the marks under all specifications and it is a skill in which many pupils find it difficult to score highly.

Supporting coursework

The wide range of maps, diagrams and skills covered in the activities in this *Teacher's Resource Book* increases pupils' understanding and experience of handling different sources of data and gives them ideas for effective methods of presenting their own data in coursework. As stated above, learning to evaluate their own answers will help with this section of their coursework.

Supporting ICT

This *Teacher's Resource Book* encourages pupils to develop their ICT skills in two main ways: using websites for further research and completing activities making effective use of ICT. For example, in Question 2 on Activity Sheet 8.5 pupils must look at a particular website to be able to answer the questions, whereas in other activities such as on Activity Sheets 3.4 and 12.5, websites for further research are suggested. Resources taken from websites are also used, as in Activity Sheets 16.3 and 16.5, where pupils might be encouraged to explore these websites and read further. Many activities suggest the use of ICT for completing the task; for example, Activity Sheet 4.5 encourages pupils to use a spreadsheet program to produce a bar graph and a pie graph.

Supporting differentiation in assessment

This *Teacher's Resource Book* is aimed predominantly at Higher Tier pupils. Nevertheless, within each activity sheet the questions are graded at different levels. Activity Sheet 11.5 on Aid, for instance, begins with shorter, simpler questions (Questions 1a–2d are all worth one or two marks each) demanding points-marking, followed by longer, levels-marked answers using more complex command words (see Questions 2e(i) and (ii), with 6 and 7 marks respectively). The extract provided as a resource on this activity sheet also contains quite high-level language more suitable for the most able pupils. However, teachers will find plenty of opportunities for differentiation within the activities and the chapter commentaries in particular highlight where activities are more suitable for lower and higher levels of ability.

British Isles

0   100 km

Europe

0   500 km

The World

4000 Miles

6000 Km

Equatorial scale

0

The World

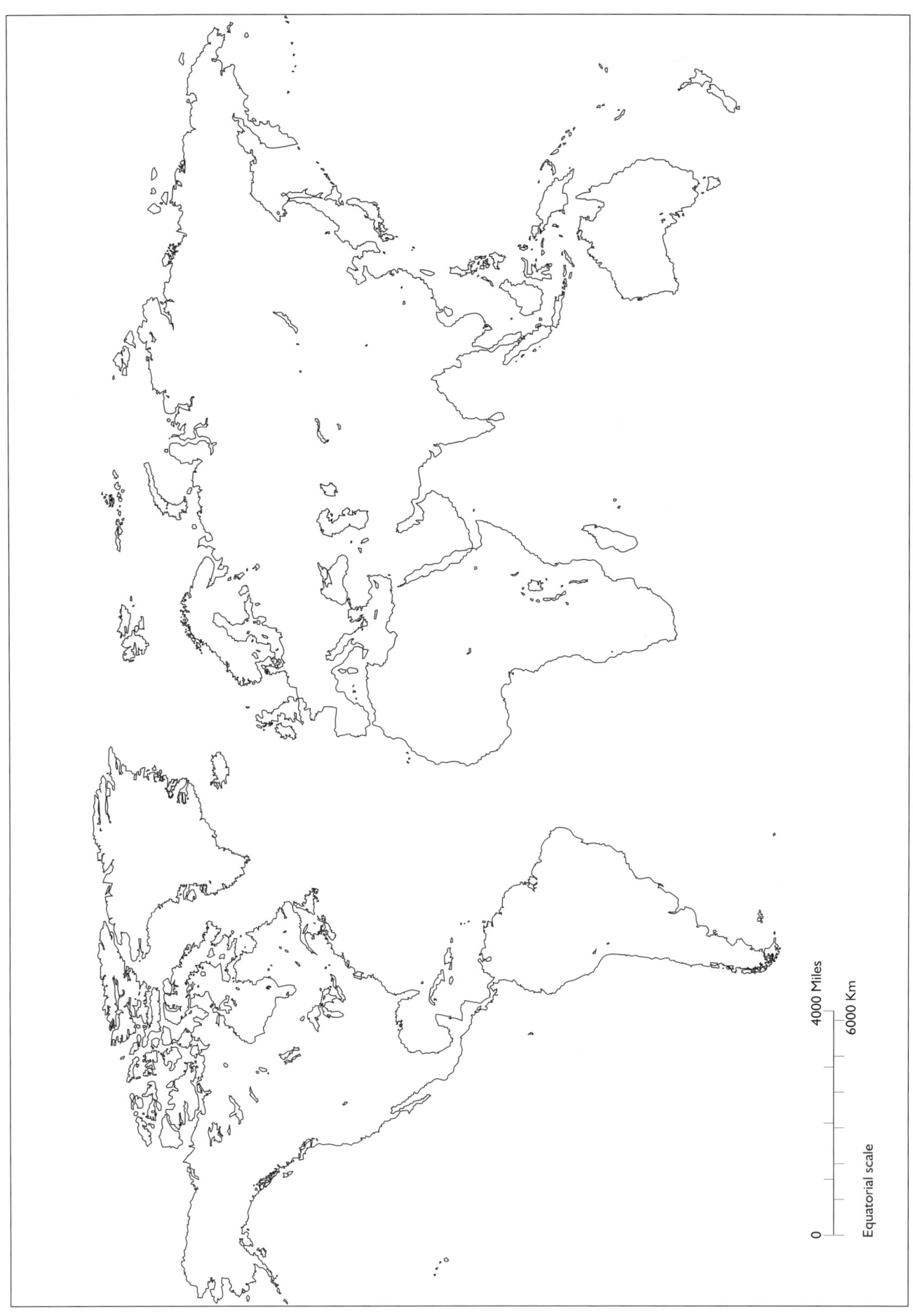

4000 Miles

6000 Km

Equatorial scale

0

The New Wider World (Second Edition): Teacher's Resource Book
Neil Punnett and Alison Rae © Nelson Thornes 2003

CHAPTER 1 contains six sections on the main GCSE and Standard Grade topics plus a case study of Brazil exemplifying some of these sections. Population is a topic most pupils find interesting, and therefore it lends itself to much discussion and debate; suggestions are included below.

Population Distribution and Density 📖 4 – 5

These two related concepts are often confused and the focus here is on differentiating between them. Factors affecting distribution and density are categorised as either physical or human. Work with population distribution and density maps allows pupils to develop their interpretation skills.

Activity Sheet 1.1 looks at both of these concepts and compares the main mapping methods for each by referring to Figure 1.1 on page 4 and a map of UK population. The questions involve interpretation and explanation of the map, using Figure 1.3 on page 4 as a source of information. The activities give pupils the opportunity to use their skills with maps and apply their knowledge of the positive and negative factors influencing distribution.

Population Growth 📖 6 – 7

The focus in this section is on the Demographic Transition Model (DTM), with reference to the recent development of Stage 5 countries shown in Figure 1.4 on page 6. The concepts of birth and death rates, and natural increase and decrease are included. The effects on society of both increasingly youthful and ageing populations could be topics for lively debate and Activity Sheets 1.2 and 1.3 could be used to form the basis of such discussion.

Activity Sheet 1.2 considers the DTM and particularly Stage 5 as seen in Figures 1.4 and 1.6 on pages 6 and 7. The questions focus on basic definitions for terms such as birth rate and require the calculation of natural population change in Germany. A tick box exercise on the effects of population growth on society and the provision of services is included.

Population Structures 📖 8 – 9

Population pyramids provide the main topic in this section. There is a general description and explanation of population pyramids and terms such as dependency ratio, with examples of calculations. Several pyramids are illustrated to show the changes in shape that occur as a country progresses through the various stages of the DTM. One excellent website for population pyramids is www.census.gov/ftp/pub/ipc/www/idbpyr.html

Activity Sheet 1.3 links population pyramids to the DTM. A pyramid is provided for Stage 4 and pupils must predict changes into Stage 5. These activities are quite theoretical and aimed at the top of the ability range.

Activity Sheet 1.4 concentrates first on one particular population pyramid, that for Kolkata (Calcutta), and a dependency ratio calculation is included. Three other pyramids, this time for individual towns in the British Isles and the Irish Republic, have to be identified by their characteristics. Group work would be a useful approach here.

Population Trends 📖 10 – 11

These pages cover changes in population growth rate, including predictions up to 2050. Factors influencing reduction in growth rate are listed on page 10. Page 11 is concerned with the contrasting experiences of MEDCs and LEDCs. The impact of AIDS, particularly in Africa, may be of particular interest to pupils.

Changing Population Structures 📖 12 – 15

Pages 12–13, on changing population structures, consider the impact on a country of having too many or too few under-15-year-olds. China's one-child policy is used as an extended example.

Pages 14–15 look at Italy and China to illustrate the issues surrounding an ageing population. Problems for the elderly themselves and for their home country are both listed in Figure 1.19 on page 14.

Activity Sheet 1.5 begins with a cartoon, which makes pupils think about total world population. This is followed by a look at population policies, providing a link with the example of China on pages 13 and 15. Both pro- and anti-natalist policies are included, with examples. There is also a consideration of the approach in the UK.

Case Study 1 Population: Brazil 📖 16 – 17

Distribution, density, structure and trends in population are illustrated in the context of Brazil. Figure 1.27 on page 17 describes the experience of one Brazilian mother, which may shock some pupils because it is so different from the family norms here.

Case Study Extra Ageing in the UK

Changes in life expectancy and the consequences of these on population structure and society are included. The very elderly are highlighted as a growing age group. Racial and regional differences in England are also discussed. This material complements Activity Sheets 1.2 and 1.3 as well as the suggested discussion of ageing.

1 Define the term *population distribution*. (2)

2 What is the most usual way of showing population distribution on a map?
 (Refer to Figure 1.1 on page 4 of the pupil book.) (1)

3 Give two advantages and two disadvantages of using this mapping method. (Hint: think
 about the type of shading and the scale used.) (4)

4 Figure A is a choropleth map, which shows population density in the UK.
 List three ways in which it is different from the map showing distribution. (3)

A UK population density.

5 Identify two areas on the map that are densely populated and another two areas
 that are sparsely populated. (4)

6 Annotate (write detailed labels on) the map to explain why the areas you have chosen are
 densely or sparsely populated. Refer to Figure 1.3 on pages 4–5 of the pupil book to
 remind yourself of possible factors, but here are some ideas to give you a start: (8)

 • Physical factors: relief, climate, natural resources, water supply.

 • Human and economic factors: transport, services (e.g. health, education), job opportunities.

Figure 1.6 on page 7 of the pupil book shows a birth rate of 9 per 1000 per year and a death rate of 10 per 1000 per year for Germany. The natural increase or decrease of a population can be calculated using the birth and death rate figures.

1 Define the terms *birth rate* and *death rate*. (4)

2 a) Write a simple formula to show how the natural change of a population, either positive (increase) or negative (decrease), can be calculated. (3)

   b) Use your formula to calculate the rate of natural change in the population of Germany. (2)

   c) Does Germany have a natural increase or decrease in population? (1)

Except in times of war or famine, most countries in the world have experienced a natural increase in population. This has been true for countries in Stages 1, 2 and 3 of the Demographic Transition Model (DTM). In Stage 4, growth has been either slight or non-existent. The situation of zero growth results in a static population. Some European populations are now in Stage 5 and are in decline, which has various consequences.

3 Complete the table below by ticking the appropriate boxes to indicate your ideas on the possible results of an increase in population for a country in Stage 3 and of a decrease in population for a country in Stage 5. The first row has been done for you. (16)

Stage 3 Country		Services provided	Stage 5 Country	
Increase	Decrease		Increase	Decrease
✓		Number of schools		✓
		Number of places in old people's homes		
		Number of job vacancies		
		Number of teachers		
		Immigration		
		Number of maternity beds in hospitals		
		Demand for the 'meals on wheels' service		
		Number of unemployed persons		

4 Give the economic and social effects of two of the changes identified for the Stage 5 country. (4)

In several MEDCs, the lower birth rate found in Stage 4 is continuing to decrease and, in some cases, has dropped significantly below the death rate. Some geographers therefore think that this new situation means we should add another section to the Demographic Transition Model (DTM), i.e. a Stage 5.

1 Figure A shows the population pyramid of a Stage 4 country. However, as a country passes from Stage 4 to Stage 5, changes in the birth and death rates will result in a differently shaped pyramid.

Label this pyramid diagram to show what you think will happen:

a) at the base of the pyramid among the 0–14-year-old groups   (2)

b) in the centre of the pyramid among the working population, aged 15–64   (2)

c) at the apex (top) of the pyramid among the 65+ age groups.   (2)

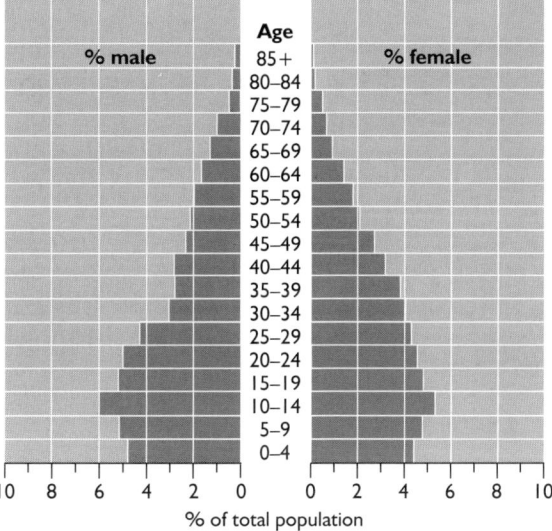

A Stage 4 population pyramid.

2 Sketch the shape of a typical population pyramid for a country in Stage 5 to show what you have suggested in Question 1.   (5)

3 Geographers use several models of which the DTM is one.
Briefly explain what is meant by the term *model*. Page 7 in the pupil book will help you.   (2)

1 Figure A shows a population pyramid for Kolkata. Migration has resulted in this pyramid being an unusual shape.

   a) Write down two characteristics that make this pyramid unusual.   *(2)*

   b) What problems are caused for the city's people by its age and gender structure?   *(4)*

   c) Given what you now know about the population of Kolkata, what can you work out about the rural areas from which people migrated to the city?   *(4)*

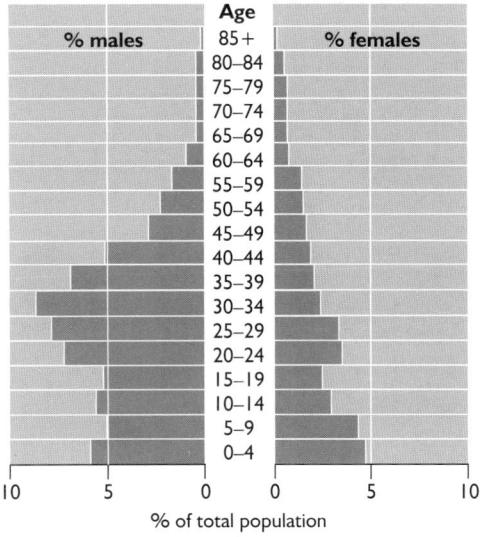

A Population pyramid for Kolkata.

2 a) Write down the equation for calculating dependency ratio.   *(2)*

   b) Work out the dependency ratio for Kolkata, using the percentages and the age categories 0–14, 15–59 and 60+ on the pyramid in Figure A.   *(5)*

3 Figure B shows population pyramids for three towns in the UK and the Irish Republic.

   a) Each pyramid has two label arrows. Choose the correct label for each arrow from the list below and write its letter in the appropriate box.   *(6)*

     A   Large proportion of elderly, especially female

     B   High birth rate

     C   Teenagers/young adults leave

     D   Low birth rate

     E   People return to bring up families in their home area

     F   Smaller middle-aged group

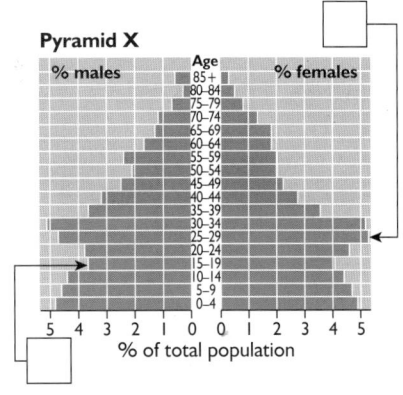 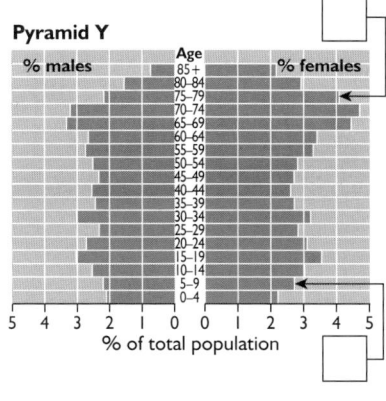 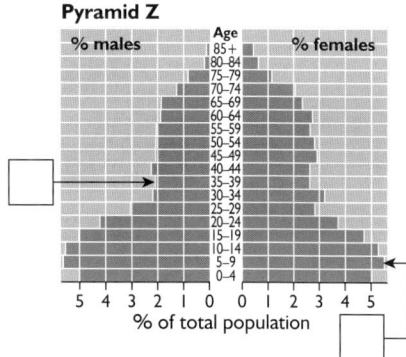

B Population pyramids for three towns in the UK and the Irish Republic.

   b) Now you know more about each pyramid, decide which town each one represents. Match the letter for each pyramid against the correct description in this list.

   • A retirement town on the south coast of England.

   • A small town in the Irish Republic, with a relatively high birth rate, but with many young adults leaving to find work.

   • A town in mid Wales that many young adults leave, but sometimes return to later.

1 This cartoon has something to say about population. Explain what you think it means. *(3)*

If a government has a pro-natalist policy, it aims to increase the country's population by encouraging people to have more children.

He's one in six billion!

Sweden is one country that has a pro-natalist policy. The government was concerned about the ageing of the population and the problems this brings. It brought in new strategies to encourage couples to have more children, including:

- longer, paid maternity leave
- paternity leave
- mother's job kept open until child is 30 months old
- cost of child care linked to parents' income.

All these make it easier to work and bring up children at the same time.

2 a) Are LEDCs or MEDCs more likely to have a pro-natalist policy? *(1)*

b) Give two reasons for your answer to a). *(2)*

c) Choose two countries from the list below that are likely to have a pro-natalist policy. *(2)*

- India
- Kenya
- France
- Sri Lanka
- Brazil
- Germany

d) What other strategies to encourage population increase could be adopted by a country? *(3)*

3 Many countries have an anti-natalist policy.

a) What is the aim of this type of policy? *(2)*

b) Are LEDCs or MEDCs more likely to have an anti-natalist policy? *(1)*

c) Choose two countries from the list above that are likely to a have an anti-natalist policy. *(2)*

d) For one country you have studied, discuss the strategies involved in an anti-natalist policy. *(6)*

4 Below is a list of possible actions a government might take to change its population. Decide whether each one is pro-natalist or anti-natalist. *(4)*

- Improve health care.
- Encourage women to work or to have careers.
- Raise the minimum age for marriage.
- Ban contraceptives.

5 Britain does not have a specific population policy, but do you think the government encourages or discourages people to have children? Give reasons for your answer. *(5)*

Ageing in the UK

In the world today one in ten people is aged 60 years or over. By 2050, this figure will be one in five. In MEDCs like the UK, this trend of ageing is even stronger. The number of older people in our population has been increasing steadily and is likely to continue to do so in the future.

The UK Census (and other similar statistics) shows how the numbers of people of pensionable age have been increasing. The term *pensionable age* refers to women aged 60 or over and men who are 65 or over.

Date of statistic	Number of people	Percentage of population
1991	10.3 million	18.2
1994	10.6 million	18.7
2001	11.2 million	19.2

In the 21st century, the number of people of pensionable age is expected to continue to rise by 1% per year until there are around 14 million people in this category by 2034. Figure A shows this expected pattern. Such growth has huge consequences for the economy. As the dependency ratio increases, funding pensions becomes more difficult.

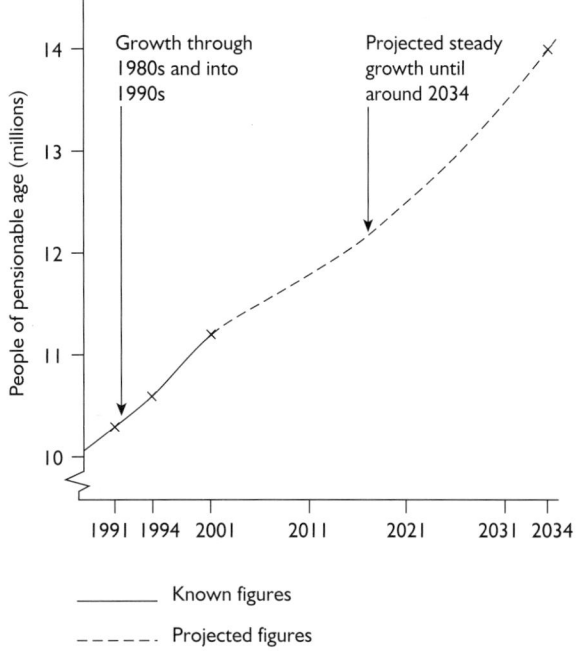

A Projected rise in the number of people of pensionable age.

Life expectancy

Life expectancy is defined as the number of years a person is expected to survive. Figure B gives an outline of how life expectancy has been increasing in the UK during the 20th century and through to the 21st. As the UK has become wealthier and the standard of living has steadily improved, people have become healthier. We eat better, spending a considerable percentage of our incomes on food, and we are more able to cure an increasing range of illnesses. These trends are likely to continue through the 21st century. Today's average life expectancy is 77 (male and female together). As women tend to live longer than men, today's average life expectancy for males is 74.9 years, but for females is 79.6 years. Some other countries have an even higher life expectancy; for example, the female life expectancy in Japan is already 80 years. Life expectancy should continue to increase, but only very slowly.

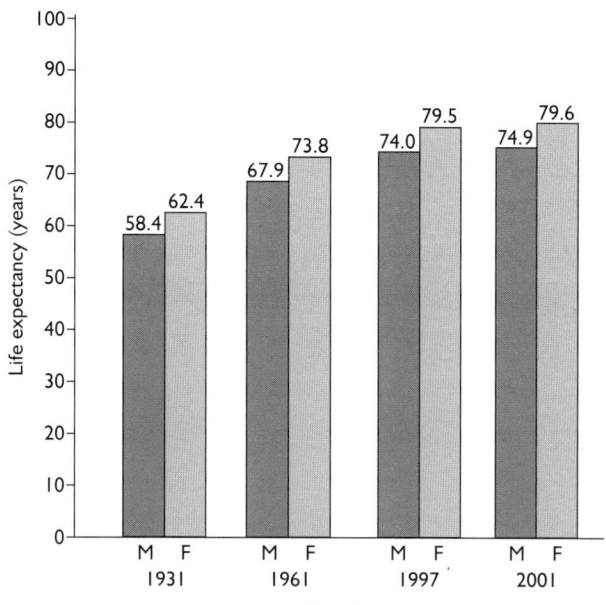

B Life expectancy in the UK.

Gender structure

One consequence of increased life expectancy is a change in the balance of the sexes in the UK population. In the 80+ age group less than 30% are male. This difference shows up clearly on the population pyramid for the UK on page 8 of the pupil book.

The very elderly

Within the older population group, it is the very elderly who are most rapidly increasing in numbers. In fact the number of people aged between 65 and 79 is likely to remain fairly constant at around 16% of the population up to 2010. However, whereas in 1961 the 80+ age group made up 1.9% of the population, by 2001 this group had risen to 4% – and this trend is expected to continue well into the 21st century. The older the age group, the faster their numbers are projected to rise. Until 2030, the 85+ group will be growing by as much as 3% a year.

Racial differences in age structure

The UK's racial minorities tend to have a smaller proportion of older people at present. Migration is normally undertaken by younger people, so many who arrived in the UK in the last few decades have not yet reached the older age group. A higher birth rate in the ethnic minorities also plays a part in lowering the average age of these groups.

Ethnic group	Percentage of population in 65+ age group
White population	16
West Indian population	3
Indian population	3
Bangladeshi/Pakistani population	1

Regional differences in age structure

Life expectancy and age structure are not the same over the whole of the UK. Figure C (below and on page 20) shows a pair of choropleth maps to show the regional differences within England. (The data for London is by borough and cannot be shown clearly on these maps.) The 'North-South divide' (i.e. the differences between northern and southern parts of England) really shows up here. For example, the male life expectancy in Manchester is only 69.9 years and is the only figure below 70 years in the country, whereas the highest male life expectancy is 76.6 years, in Cambridgeshire. The variation in female life expectancy is similar, from 75.5 years in West Yorkshire to 81.4 years in Dorset and in Bromley, Kent.

80+ years
75–79 years
70–74 years
<70 years

C (i) Regional differences in male life expectancy in England, 2001.

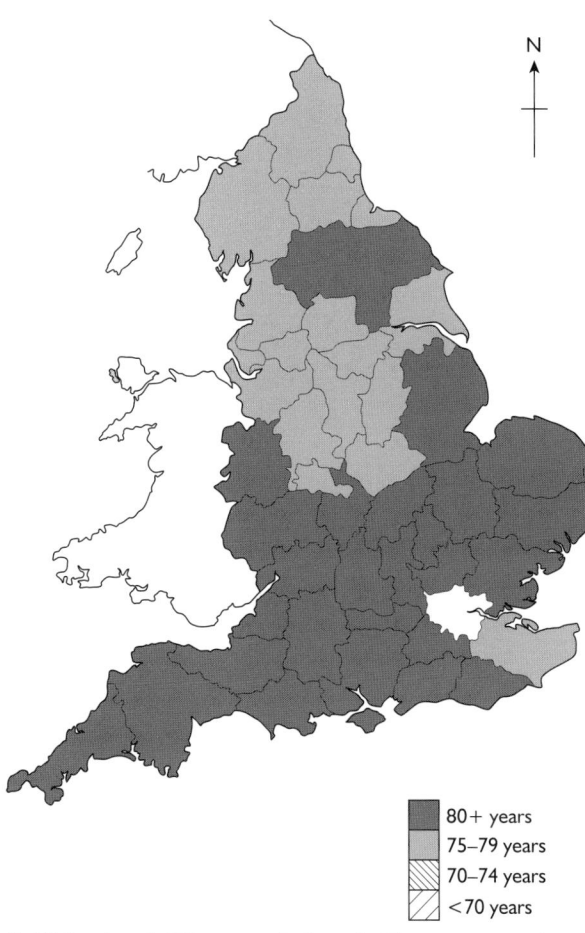

N

Perhaps the most worrying aspect of these statistics is that the differences between the wealthiest and the poorest regions are increasing all the time. Regional variations are wider now than ten years ago, even though the overall averages for life expectancy and for standards of living have both increased. No region in the life expectancy top ten is north of Birmingham and only three inner London boroughs have figures as low as those in the north. If people are disadvantaged during their lives, they will not live as long as those who are more advantaged. Disadvantage occurs in many respects, for example:

- quality of food
- type of employment
- housing conditions
- level of income.

80+ years
75–79 years
70–74 years
<70 years

C (ii) Regional differences in female life expectancy in England, 2001.

Exam Practice Questions

1 a) Draw an outline population pyramid for the UK and annotate it to show the characteristics of the apex for both genders. *(5)*

b) Explain, either by further annotation of your population pyramid or in a paragraph, what is likely to be happening at the base of the UK population pyramid at the present time. *(6)*

2 a) Using Figure C, describe the pattern of life expectancy over England. *(6)*

b) Suggest how the government could try to level out the regional differences in life expectancy by solving the problems of inequality. *(3)*

Ageing in the UK

These model answers are designed to show you how to get full marks for some of the longer questions in the Case Study Extra for this chapter. Read the question carefully and write your own answer. Then read the model answer and the examiner's notes to see what he or she is looking for when awarding the marks. Decide how many marks you think the examiner would give your answer. Decide how to change your answer to this particular question to increase your marks. What will you do differently next time you answer a similar question?

1 a) Draw an outline population pyramid for the UK and annotate it to show the characteristics of the apex for both genders.   (5)

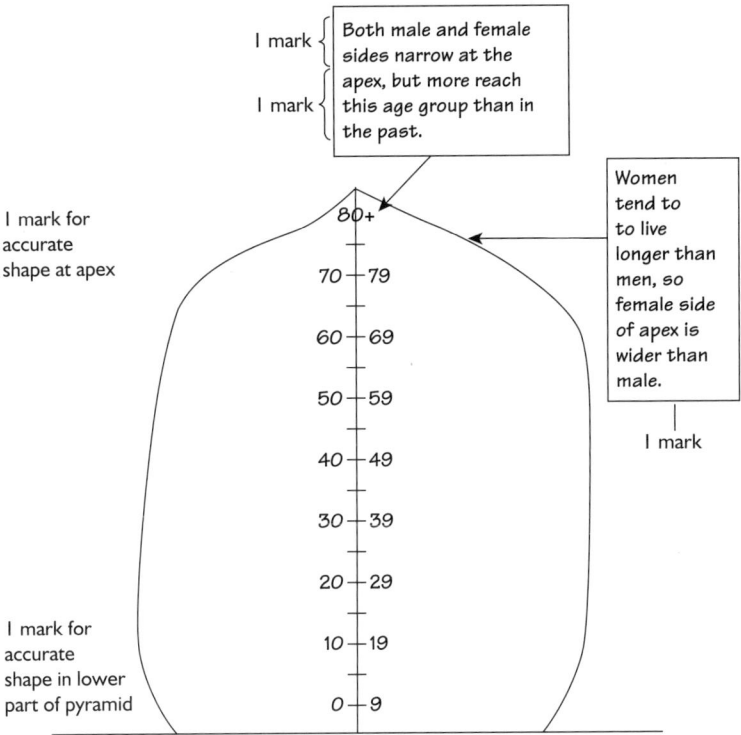

I mark { Both male and female sides narrow at the apex, but more reach
I mark { this age group than in the past.

I mark for accurate shape at apex

Women tend to to live longer than men, so female side of apex is wider than male.

I mark

I mark for accurate shape in lower part of pyramid

80+
70 — 79
60 — 69
50 — 59
40 — 49
30 — 39
20 — 29
10 — 19
0 — 9

Improve Your Mark!

Annotation is a very important skill and if you can learn to do it well you can guarantee yourself maximum marks on these questions. Basically, annotations are detailed labels. They contain information and so are longer than normal labels.

This is not a 'levels' question, but is 'point marked'. This means you gain one mark for each thing you do well. Here, the five marks are awarded as follows:

1 mark for the shape of the apex (top) of your pyramid. It must curve in and be narrower on the male side than on the female side.

1 mark for the shape of the rest of the pyramid. It must narrow at the base to reflect our low birth rate, with male and female sides equal.

3 x **1 mark** for the labels at the apex of the pyramid, as follows:

• both male and female sides narrow at the apex

• more people reach this age group than would have in the recent past

• female side is wider than male side due to longer life expectancy.

b) Explain, either by further annotation of your population pyramid or in a paragraph, what is likely to be happening at the base of the UK population pyramid at the present time.

(6)

> The width of the lower part of a population pyramid is controlled by the birth rate in that country. In the UK at present the birth rate is relatively low and has been decreasing. Fewer children are therefore born each year. There is no difference in the width of the male and female sides of the pyramid in this younger age group, as is the case at the apex. Almost the same numbers of boys and girls are born, and infant mortality rates are very low in an MEDC such as the UK.
>
> There are various reasons for the fall in the birth rate.
>
> - Economic development encourages a lower birth rate. People have money to spend and desire the luxuries in life. Children are a heavy cost and they usually do not earn for themselves until they are eighteen or older.
>
> - An increasing proportion of women are career orientated. This may lead them to postpone having children until they are older and therefore they tend to have fewer children. More families today have an only child. An increasing number of couples decide not to have children at all.

Improve Your Mark!

The first paragraph describes the shape of the pyramid at its base and begins to give some basic explanation for this, i.e. that the birth rate has declined. The second paragraph then goes into greater detail to explain reasons for the drop in the birth rate.

This is a 'levels' question. The mark allocation and difference between the levels for this particular question are as follows:

Level 1 (1–2 marks): *For a simple description of the shape of the pyramid; little explanation would be offered.*

Level 2 (3–4 marks): *For a good description plus some explanation, i.e. that the birth rate has declined.*

Level 3 (5–6 marks): *For a good description plus a higher level of explanation, including reasons for the decline in the birth rate, as shown in the model answer.*

2 MIGRATION

CHAPTER 2 contains five sections tackling various topics under the umbrella of migration, plus a case study on migration to California. The reasons why people migrate underlie the whole of the chapter.

What is Migration? 📖 22

This section introduces the basic concepts and terms used in the study of migration. Some of the world's major migration routes are mapped in Figure 2.3 and this could be supported by the use of an atlas to see the exact locations of the countries involved.

Activity Sheet 2.1 supports the definition of terms. Pupils are asked to calculate a simple net migration to check their understanding of the concept. They must categorise particular types of migration, such as internal, international and economic migration, and these can also be marked on the world map on page 10 of the *Teacher's Resource Book*, in a similar way to those in Figure 2.3 on page 22.

Refugees 📖 23

A whole page is given to this particular type of migration. Pupils will have heard news referring to asylum seekers and refugees; the terms are not always used correctly and are often interchanged. An opportunity therefore exists for discussion and clarification of the causes of having to seek refuge and the consequences. Useful websites on this topic include www.refugees.org, www.unhcr.ch and www.refugeesinternational.org

Activity Sheet 2.2 also considers the issues surrounding refugees, highlighting the differences between the LEDC and MEDC experience, the factors creating refugees and methods of coping with them. No matter when you are using this activity sheet there will doubtless be some current refugee crisis in the world. News reports and newspaper extracts could be utilised to generate class discussion or debate.

Migration into the UK 📖 24 – 25

Historical description and some explanation of the various migration streams into the UK are provided. Concentrations of ethnic groups are mapped in Figure 2.7 on page 24, which gives an opportunity to consider the use of proportional symbols, a higher-level presentation skill if used in pupil coursework. The problems faced by UK immigrants are also discussed. Pupils might try to put themselves in the shoes of others. Ethnic minority children could explain their own families' experiences to the group.

Migration within the UK 📖 26 – 27

Page 26 tackles push and pull factors in detail, set against the background of internal migration within the UK. Change in London is particularly highlighted. Figures 2.11 and 2.12 on page 26 are choropleth maps showing population change, providing an opportunity to consider this mapping method. Pupils could discuss their family moves and Activity Sheet 2.5 is useful here.

Activity Sheet 2.3 also looks at push and pull factors, and considers how the same situation can affect people in different circumstances. The idea of neutral factors, which do not influence individuals, is also introduced.

Migrant workers 📖 28 – 29

These pages involve a case study on the Turkish community in Germany. There are opportunities for pyramid interpretation and more work on push and pull factors.

Activity Sheet 2.4 includes details of a Turkish village and so complements the case study above. Push and pull factors and the consequences of this type of migration for the country of origin, as well as for the destination, are considered.

Activity Sheet 2.5 includes questions that support the sections on migrant workers on pages 28–29 and on migration within a country on pages 26–27 of the pupil book, this time in the context of Malaysia. This allows comparison of the LEDC and MEDC situations. Pie chart construction and proportional flow arrows provide useful skills practice.

Case Study 2 Immigrants into California

Higher-level writing skills can be practised as this material can be included in a levels answer. The idea of illegal immigration is introduced, as well as the unevenness of the distribution of the migrant population.

Case Study Extra Migration to France 1954–2003

In France today, many people perceive an immigration problem. The sources and destinations of migrants are discussed, as well as the reasons for settling, status in society, living standards and current attitudes towards the migrants.

1 Write a definition for each of the migration terms in the diagram below. Page 22 in the pupil book will give you a start. A geographical dictionary may also help you. *(7)*

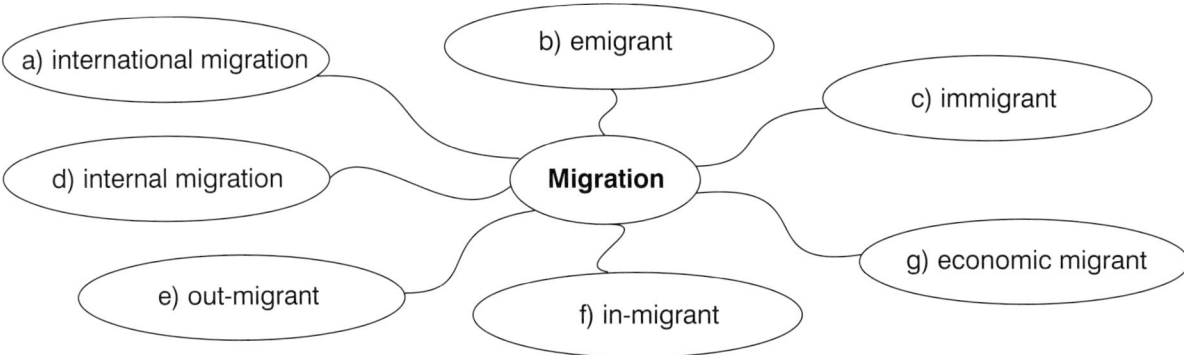

a) international migration
b) emigrant
c) immigrant
d) internal migration
Migration
e) out-migrant
f) in-migrant
g) economic migrant

2 If a country had 645 010 immigrants and 239 803 emigrants in one year, calculate its net migration for that year. *(2)*

3 Decide whether each of the situations set out below can be classified as internal, international or economic migration. Put a tick in the appropriate box. *(4)*

Household	Internal	International	Economic
A A third child was born so this German family moved into a larger house in the same district of Bonn.			
B A retired couple from the UK sold their home in York and moved to the Dordogne region in France.			
C A sick miner and his family from Minas Gerais State in Brazil moved to São Paulo in the hope of finding factory work that would not cause health problems for the father.			
D A farm labourer from Zimbabwe, displaced by the government-enforced eviction of his employer, went to South Africa to find work on a farm there. His family remained in Zimbabwe, hoping he would be able to send money for them to live on.			

4 On an outline world map, draw three arrows to represent migrations B, C and D above. Label each arrow with information about the relevant type of migration. For migration A, just add a label in the correct area of the map; no arrow is needed. *(7)*

1 Write down a full definition of the word *refugee*. (3)

2 List four countries which have produced large numbers of refugees. (4)

3 Which of the following situations are likely to result in refugees? Tick the appropriate boxes. (5)

Situation	Yes	No
A volcanic eruption		
People looking for new jobs		
Civil war		
War with another country		
Tourism		

4 a) At any one time, there are at least 20 million refugees in the world. Do the majority of them seek refuge in LEDCs or in MEDCs? (1)

　b) Give reasons for your answer. (3)

5 Study the extract on refugees' needs below.

　Imagine you are in charge of the planning of a charity campaign based in an MEDC to help a new group of refugees in an LEDC who have fled from a recent crisis.

　a) Write a paragraph to explain how you would plan to cope with the refugees' immediate needs. (7)

　b) Design a poster to advertise your campaign. (10)

Refugees' needs
Refugees usually need immediate support in terms of food, shelter, clothing and medical care. But they also need protection and the security of knowing they will not be sent back to the country they have left or be separated from their families. In the longer term they need support for education, health care, employment and a chance to be responsible for their own lives.

From **The Third World** by R.Beddis.

6 What problems might be created for less-developed countries by the arrival of refugees from a neighbouring country? The extract on refugees in Africa will help you to answer this question. (5)

Refugees in Africa
Almost half the world's refugees are in southern and east Africa. Poor countries find it hard to cope with the arrival of large numbers of refugees when many of their own people are struggling to survive. Many African countries are nevertheless generous to refugees…

From **The Third World** by R.Beddis.

7 Many people assume that all refugees lack skills and education. In fact, some are highly skilled. The Ugandan Asian population, which came to the UK in 1972, included business people, doctors, etc. Discuss and note down the ways in which refugees might benefit their host country. (5)

1 Define the terms *push factors* and *pull factors*.   (4)

2 Some of the factors set out below are pushes and some are pulls. Draw a line to link each factor to either push or pull factors as appropriate.   (15)

Job prospects	Natural disasters	Low income	Housing shortages	Health care
High wages				High standard of living
		Pull factors		
Improved housing				Tolerance
		Push factors		
Political or social unrest				High unemployment
Intolerance	Educational opportunities	Adverse climatic conditions		Attractive environments

3 The table below shows an example of push and pull factors, and how they apply to different types of people and affect their decisions on whether to migrate. Not all people react in the same way to the same set of pushes and pulls. This is called differential migration and involves the following factors:

- negative (push) factors
- positive (pull) factors
- neutral factors, which do not affect the person's decision.

	Household A		Household B	
	Single, middle-aged man, who likes fishing		Poor young couple with children; father hates fishing	
Place of origin (where they live now)	Little work	–	Little work	
	Fishing river		Fishing river	0
Place of destination (place to which they might migrate)	Factory jobs for men available		Factory jobs for men available	
	Factory jobs for women available		Factory jobs for women available	+

a) Fill in the blanks in the table to show whether you think the migration factors are positive (+), negative (–) or neutral (0) for each household. Three boxes have been filled in for you.   (5)

b) Decide which household would be more likely to migrate. Give reasons for your answer.   (3)

Sakaltutan is a village in central Turkey with about 1000 inhabitants. It has been a poor settlement, dependent on agriculture, for a long time. With a high birth rate and limited resources, it soon became overpopulated. Because of this, there were not enough jobs for the men to do. There was only so much land that could be farmed and the demand for craftsmen was limited.

When an all-weather road was built, providing a more reliable link between Sakaltutan and other settlements, surplus produce could be sold in the region's towns. Some farmers therefore had the money to buy machines, although this also led to job losses for some farm labourers.

The education system in Turkey steadily improved during the 1970s and 1980s. This led to a growth in aspirations. Fewer young people were satisfied with the standard of living offered in Sakaltutan and hoped they could do better elsewhere.

Outward migration occurred. People went to:
- Adana – the nearest city
- Ankara – Turkey's capital
- Germany (see pages 28–29 in the pupil book).

1 Define the term *overpopulation*. (2)

2 Explain briefly how improvements in education encourage people to migrate from villages such as Sakaltutan. (3)

3 There are advantages and disadvantages for both the places of origin and destination in any migration.

a) The following list gives some of the consequences of the economic migration from Sakaltutan to Germany. In the table below, indicate which consequence would be an advantage or a disadvantage in each location. The first two have been done for you. (9)

A Pressure reduced on jobs and resources.

B People of working age leave.

C Money may be sent back to the village.

D Cheaper labour available, especially for the dirty, unskilled jobs.

E Migrants feel discriminated against.

F Mainly males migrate and divide families.

G Cultural advantages, e.g. new foods, music, etc.

H Migrants may be a drain on local services.

I Migrants form an ethnic group that does not mix.

J An elderly population is left.

K People with skills and education are the most likely to leave.

Location	Advantages	Disadvantages
Sakaltutan	A,	B,
Germany		

b) Copy the table above, but insert 'Other parts of Turkey' instead of 'Germany'. Complete this second table. (11)

Malaysia is an NIC, or Newly Industrialising Country. Most people live in the Peninsula, the mainland section of the country. Its economy is growing quickly and, because of this, some regions of the country have many more opportunities than others. About 5% of the population migrates each year.

1 Use the data in the table below to draw a simple pie chart showing the relative proportions of the different categories of migration. *(4)*

Type of migration	% of total migrants in 1990
Intra-state (within the same state)	51.6
Inter-state (between states)	40.0
Emigrants (leaving the country)	8.4

From Department of Statistics, Malaysia.

2 a) Based on the information in your pie chart, are more migrations short distance (intra-state) or long distance (inter-state and emigration)? *(1)*

  b) Give three reasons for your answer. *(3)*

3 Flow arrows are often the best method of showing migrations. The width of the arrow is proportional to the value it represents.

  Table A on page 29 shows internal migration for selected states in Malaysia in 1990. Using a scale of 1 mm to 5000 migrants, draw flow arrows to represent these migrations on Figure B on page 29. Kuala Lumpur has been done for you. *(8)*

4 All five states in Table A on page 29 have both in-migration and out-migration. Explain why this can occur. *(6)*

Name of state	Key pull factor	In-migrants	Out-migrants
Kuala Lumpur	Capital city	22 069	51 669
Selangor	Area surrounding capital	89 007	62 866
Pinang	Outer development area	54 927	48 054
Pahang	Have government sponsored rural development projects	59 456	59 745
Kedah		68 108	62 430

A In-migration and out-migration in Malaysia.

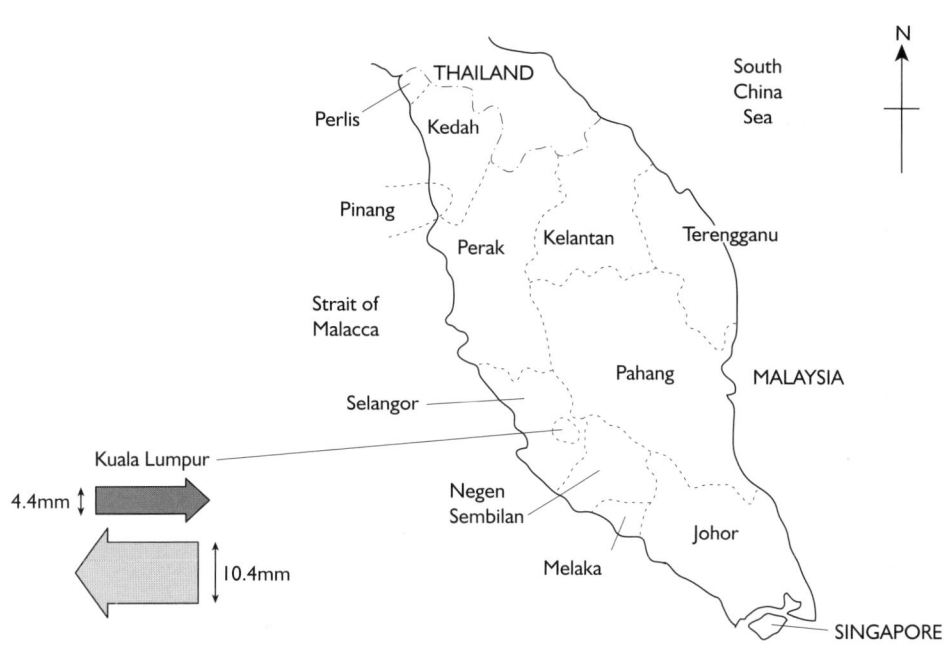

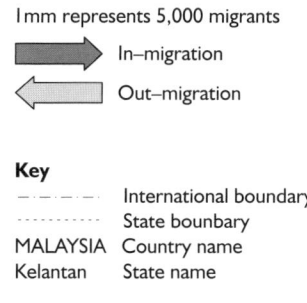

Scale of arrows
1mm represents 5,000 migrants

In–migration

Out–migration

Key

— · — · — International boundary
· · · · · · · · · State bounbary
MALAYSIA Country name
Kelantan State name

B Map showing in-migration and out-migration in Malaysia.

Migration to France 1954–2003

France has been a desirable destination for migrants for a long time. In the 19th century, France actively colonised large parts of what are now the LEDCs, particularly in Africa, in much the same way that Britain did. Most French colonies were located in West Africa (e.g. Cote d'Ivoire or Ivory Coast) and North Africa (e.g. Morocco and Algeria).

Number of immigrants and their places of origin

Over the years, the number of immigrants living in France has increased substantially and the balance of their places of origin has changed (see Figure A). This change has been more rapid since 1954. Migrants and their children, some of whom were born after their parents' arrival in France, grew steadily from 1.7 million in 1954 to 4.3 million in 1999. This trend is likely to continue. Migrants to France originate from three main areas.

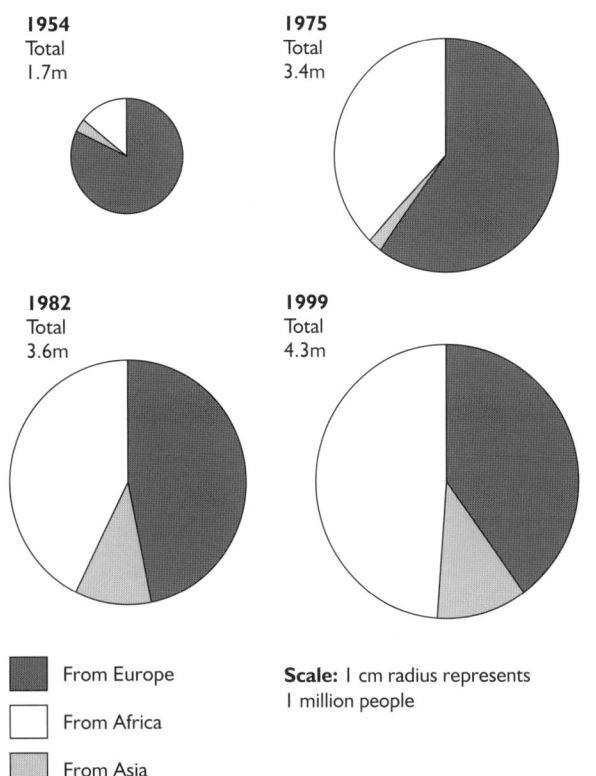

1954 Total 1.7m

1975 Total 3.4m

1982 Total 3.6m

1999 Total 4.3m

■ From Europe
□ From Africa
▨ From Asia

Scale: 1 cm radius represents 1 million people

A Number of immigrants in France and the change in their places of origin.

Migrants from other parts of Europe

In 1954 a large majority of migrants (84%) were from other parts of Europe and mainly from neighbours, such as Spain and Italy. The EU now allows its citizens to settle and work freely in any of the member countries, but there have been considerable flows of migrants between European neighbours for a long time. By 1990, only 41.3% of migrants settled in France had a European origin, a significant decrease proportionally. However, the actual numbers of migrants from within Europe have not decreased.

Migrants from Africa

The people of African origin who now live in France are a mixture of black West Africans and Arab North Africans. West Africans tend to be present in smaller numbers and have usually been a part of French society for longer. It is the North African group which has become much more dominant over recent years and has almost certainly been the more controversial. The total numbers of African migrants have increased from 229 500 (13.5% of the immigrant population in 1954) to 1 965 600 (46.8% in 1990).

Migrants from Asia

Asian immigration has also increased from being an almost negligible 42 500 (2.5% of immigrants) in 1954, but increasing to 499 800 (11.9% of the immigrant population) by 1990. This is almost a twelve-fold increase over that time.

Returning migrants

A counter-stream (or counter-current) consists of those migrants who return to their origin due to some dissatisfaction with the destination. All examples and types of migration contain a proportion of people who return. In France during recent decades, one in ten migrants has returned home, including some who, having gained education and experience in France, return to their home country to use what they have learned. Others find themselves disappointed with their new life or miss their family, friends and culture too much.

Status of migrants

Migrants have entered France for all sorts of reasons, including political, economic and family reasons.

Political reasons

In 1990, 121 000 people were classed as political refugees and more, claiming to be in this category, come every year. In 1997, 21 000 new migrants asked for political asylum, although only 4100 were granted this status. Most of the remainder were deported.

Economic reasons

The majority of migrants in the world today, and indeed in the past, are economic migrants, i.e. they are seeking a better quality of life. Because MEDCs are not willing to accept all who wish to improve their opportunities in this way, some migrants try to claim their lives are in danger, in the hope that this will result in them being allowed a new home.

Family reasons

In 1997 for example, 15 500 people went to France to join members of their families already living there, and on whom they claimed they were dependent. Of these, 7100 were children.

French nationality

Some migrants have been given French nationality, but others have not. It has depended on their circumstances and background. In 1997, 21 000 people got French nationality by marrying a French citizen and 60 000 others gained it with permission from the government. Many of the latter (32 500) were children born in France to foreign (i.e. migrant) parents. Foreigners asking for French nationality are generally younger and better educated than the average native French population. Of those who were given French nationality in the 1990s, one third had had higher education. Only 37% were unqualified, compared to 45% of the native French population.

Distribution of the migrant population in France

Eighty percent of migrant households have now lived in France for ten years or more, but they are not evenly spread throughout the country.

Figure B shows that 68% of migrants are classified as urban, compared with only 38% of the non-migrant French population. Most migrants feel that they are more likely to get a job in a town or city. Also, urban areas are more mixed culturally and linguistically, so acceptance of migrants is more likely.

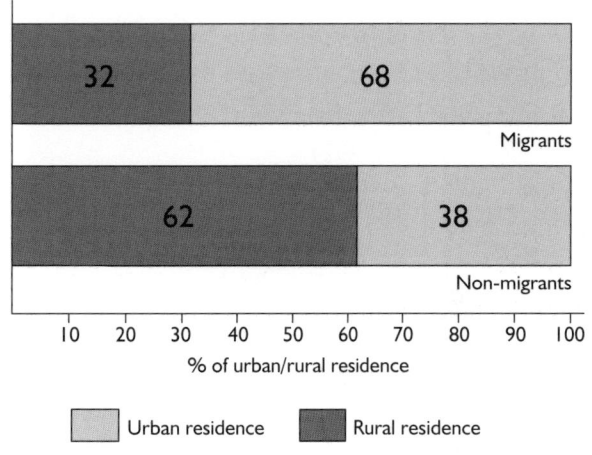

B Distribution of the migrant population in rural and urban areas.

Migrants are not evenly spread in terms of regions (see Figure C) either. Forty percent of all migrants have settled in the Greater Paris area. Brittany, in the north-west of the country, classes less than 1% of its population as migrant.

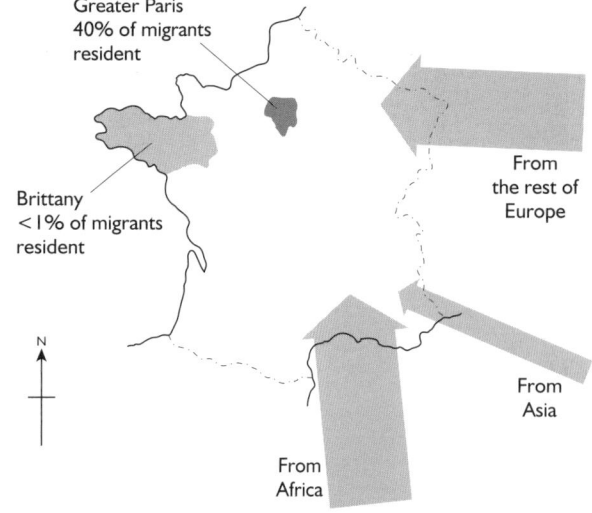

Scale of arrows: 1mm width represents 2% of total immigrant population
C Distribution of migrant population by place of origin in 1990.

Employment and living standards

Although the average migrant is relatively well educated, a higher proportion of them do manual work than the native French population. A smaller proportion of migrants are office workers or professionals than the norm. This is partly because many migrants are Moslem, a culture in which women are less likely to work outside the home. Women of Moroccan origin have the lowest employment level.

Because they are poorer, a greater proportion of migrant households live in public housing than do the native population. Traditionally they have larger families, which makes overcrowding worse, although over time migrants tend to reduce their birth rate to the level of the host country.

Attitudes towards migrants in France

A poll taken in March 2000 showed considerable feeling against the migrant communities in France. The results are shown in the table opposite .

Racist incidents have increased at football matches. The Paris St. Germain team have had to launch an anti-racist campaign.

In 1998, France won the World Cup with a multiracial team yet in the following presidential election 15% of the first round vote went to Jean-Marie Le Pen, the National Front candidate, showing the strength of anti-immigration feeling. The main criticism of migrants is that they refuse to adapt to the French way of life. However, many native French people are tolerant, believing there should be no discrimination. Greatest tolerance is found in the media, among young people and those who are more highly educated.

% of respondents	Opinion
61	Too many foreigners in France
63	Too many Arabs in France
32	France should stop taking migrants
52	Immigration is the main cause of crime
71	Migrants take advantage of the social security system
57	Migrants get priority in medical care
52	Migrants get priority in housing

Exam Practice Questions

1 Why are proportional circles, such as those used in Figure A on page 30, useful for showing population change over time? *(2)*

2 Describe the changes in the places of origin of migrants to France since 1954. Use Figure A to help you. *(5)*

3 a) Define the term *counter-stream (counter-current)*. *(1)*

   b) Why do all examples or types of migration have a counter-stream? *(2)*

4 Describe and explain the main pull factors of a country such as France for potential migrants. *(8)*

Migration to France 1954–2003

These model answers are designed to show you how to get full marks for some of the longer questions in the Case Study Extra for this section. Read the question carefully and write your own answer. Then read the model answer and the examiner's notes to see what he or she is looking for when awarding the marks. Decide how many marks you think the examiner would give your answer. Decide how to change your answer to this particular question to increase your marks. What will you do differently next time you answer a similar question?

2 Describe the changes in the places of origin of migrants to France since 1954. Use Figure A to help you. (5)

> The four pie charts increase in size considerably, showing that the number of immigrants living in France has increased from 1.7 million people in 1954 to 4.3 million in 1999. The sectors into which each pie is divided also change in proportion. In 1954, people from other parts of Europe made up the majority, around 80%. By 1999, this group made up only about 40% of migrants. In the same period, the numbers of Africans grew from under 20% to almost half and Asians from 2.5% to over 10%. Since the total numbers of immigrants to France have increased as well, this represents a large growth in the numbers originally from Africa and Asia.

Improve Your Mark!

The main place to find information is in the pie charts in Figure A. You need to comment on both the size of the circles, which show the number of migrants, and on the changes in the three sections within each circle. For the best answer, refer to particular diagrams and give figures.

Level 1 (1–3 marks): *To reach this lower level you would have referred to only some of the pie charts and not in sufficient detail.*

Level 2 (4–5 marks): *You would have referred to all four diagrams and quoted figures from them.*

4 Describe and explain the main pull factors of a country such as France for potential migrants. (8)

> Some migrants arrive in France as political refugees. These are people whose lives are threatened in their own country. A democratic country like France offers such people safety. Once in France people will be safe politically and can practise their religion freely.
>
> Many people in the world desire a better standard of living than the one they have at present. Most are willing to work for this, but need the opportunities to be able to do so. MEDCs offer more chances than LEDCs, so people are attracted to migrate. State benefits and help with housing are also available in times of need.
>
> Educational places, better health care and more job opportunities are real pull factors. Most people wish to give their children opportunities they did not have. Education is seen by migrants as the key to a better life. Many will have come from countries, especially in Africa, where access to medical services is poor and there are large numbers of people per doctor.

For those who already have close family living in France, the appeal of joining them, and so improving their lot, is also understandable. In 1997, 15 500 such people arrived in France. Some children arrive unaccompanied to join existing family members in France.

In the 19th century, France was one of a few European countries which had several colonies, countries which today are LEDCs. There is therefore a particular pull for the people in these countries (e.g. Ivory Coast and Algeria) to come to France. Many already speak French, which makes settling into the new country much easier.

Paris, as the largest city in Europe, also has a particular attraction for migrants. Its more multicultural city life and greater number of job opportunities make it a likely choice.

Improve Your Mark!

France's pull factors can be found in various parts of the Case Study Extra and extracting material like this is a skill you need to master to obtain the highest level marks on the longer answers. The information then needs to be organised in a sensible order, so that related points support each other and the links are clear. For top marks, always include points which are particularly relevant to the case study. Don't just make comments on pull factors that could apply to any MEDC. The points about France being a colonial country and about Paris are the key ones here.

Level 1 (1–3 marks): *This answer would have suggested only a few pull factors with little explanation as to why these factors would be attractive to people.*

Level 2 (4–6 marks): *In this case, more pull factors would be given with a moderate amount of additional material.*

Level 3 (7–8 marks): *For this level you would give a full list of factors, well explained, as in the sample answer, in which each main point is made and then explained.*

CHAPTER 3 deals with settlement and urban growth. This is covered under five sections: site and situation, hierarchies, urban land use models, traditional land use in the CBD (central business district) and old inner city areas, and residential environments in British cities. The final section investigates New York as a case study of a city in a developed country.

Site and Situation 📖 34 – 35

The location factors for settlements can be applied at a variety of scales, from hamlets to cities. Activity Sheet 3.1 focuses on some of the most important factors.

Activity Sheet 3.1 focuses on the siting of settlements. A sketch map of Bickleigh in Devon is used to highlight important siting factors. Pupils are then asked to draw sketch maps of a fishing village in south-west England from a map provided. Finally, pupils tackle a challenging exercise to draw an annotated sketch map of the village, town or city in which they live.

Patterns 📖 36

This topic is especially suited to OS map work. Almost any area of the UK will reveal examples of dispersed, nucleated or linear settlement. Lincolnshire and Norfolk are particularly fruitful areas, often showing all three patterns within a small area.

Activity Sheet 3.2 uses the OS map of East Yorkshire on page 41 of the pupil book and requires the pupils to draw annotated sketch maps of three named settlements and then to identify the pattern of settlement in each case. A second question uses a map of Bridgwater in Somerset to investigate the changing functions of larger settlements. Originally an important river port and market town, Bridgwater is now a growing industrial centre.

Functions 📖 37

A good exercise here is to use the idea of Figure 3.11 on page 37, but using examples from the pupils' home region. This can lead to useful discussion on changing functions.

Hierarchies 📖 38 – 39

The three different methods used to determine hierarchies are clearly described. Figure 3.13 on page 38 can be used again to name examples from the pupils' home region. Figure 3.15 on page 39 introduces Central Place Theory. At this stage, the more able students could be introduced briefly to Christaller's work. Activity Sheets 3.3 and 3.4 focus on the range and number of services in settlements of differing sizes. Activity Sheet 3.4 requires pupils to access internet websites and will make a good homework exercise.

Activity Sheet 3.3 considers settlement hierarchy using a table that shows six shops and services found in eight neighbouring settlements in a rural area of eastern England. The questions examine the threshold population and the range of goods and services. The pupils are also required to complete a scattergraph.

Activity Sheet 3.4 examines the settlement hierarchy further using the example of Beverley in East Yorkshire. Pupils have to answer questions using a flow-line map and construct two flow lines themselves. This activity sheet also includes a question requiring use of the internet to find information about contrasting shopping centres in York and Hull.

Urban Land Use Models 📖 42 – 47

The pupils' local town or city can make a useful case study. Try to apply the models of Burgess and Hoyt to it. The characteristics of suburbia can be studied; see how many of the features described on pages 46 and 49 pupils can recognise.

Activity Sheet 3.5 is concerned with urban land use and includes questions on graph interpretation and Hoyt's Sector Model.

Case Study 3 New York 📖 50 – 52

This case study begins with a map and description of Manhattan Island, the original site of New York. The main focus of the case study is on problems resulting from the growth of the city, including high land costs, urban decay, immigration, traffic congestion, unemployment, crime, pollution, water supply and climate. Figures 3.42 and 3.43 on pages 51 and 52 are especially useful in showing the centrifugal movement of immigrants within the city and the vicious circle operating within the ghetto. Useful alternative case studies include the Randstad and Paris.

The Randstad is the name given to the multi-nuclei agglomeration of five million people in the western Netherlands. The Randstad forms a giant horseshoe of urbanisation running from Dordrecht in the south-east through Rotterdam to the Hague, swinging northwards along the coast through Leiden and Haarlem to Ijmuiden and then running eastwards through Amsterdam to Utrecht. Nearly 40% of the Dutch population live in the Randstad. Dutch planners have introduced planning controls to preserve the remaining countryside between the main cities – 'green wedges' – and also to preserve the countryside within the horseshoe-shaped conurbation – the 'green heart'. Growth is directed outwards into corridors and encouraged to locate in the less prosperous areas of the country such as the north-east.

Paris experienced urban sprawl during the 1920s and 1930s, in a similar way to London. Rather than following London's example of preventing further urban sprawl by creating a green belt, the planners in Paris directed it into two corridors of growth running through Paris from south-east to north-west. Five new towns including Evry and Marne-La-Vallee were developed within these corridors. Extensive urban redevelopment has taken place within Paris' inner suburbs, including the La Defense office, commercial and residential complex with its futuristic architecture. Further information can be found in the Case Study Extra in Chapter 4 of this *Teacher Resource Book*.

Case Study Extra The Growth of London

New York is a comparatively recent city. Many European cities are very much older and their extended evolution has resulted in a more varied urban structure. London is used as an example of the extended period of growth that many European cities demonstrate. The importance of town and regional planning during the 20th century is emphasised. Pupils are required to draw a line graph showing London's population growth and to comment on population density curves for the city.

The exact locations or sites of hamlets and villages were carefully chosen by the original settlers. In Britain the original settlers would have been either Celts, Romans or Anglo-Saxons. In those times basic needs for survival such as sources of water, food, fuel and shelter were most important. It was often vital to choose a dry site that could be defended against possible attack during times of war. In more peaceful times, route centres (nodal points) and bridging points gave a site significant advantages.

1 Figure A is a sketch map showing the site of Bickleigh, a village south of Tiverton in Devon. The numbers highlight five factors that influenced the choice of this site.

a) Name each factor. (5)

b) For each factor, write a sentence explaining its importance in the choice of site. (5)

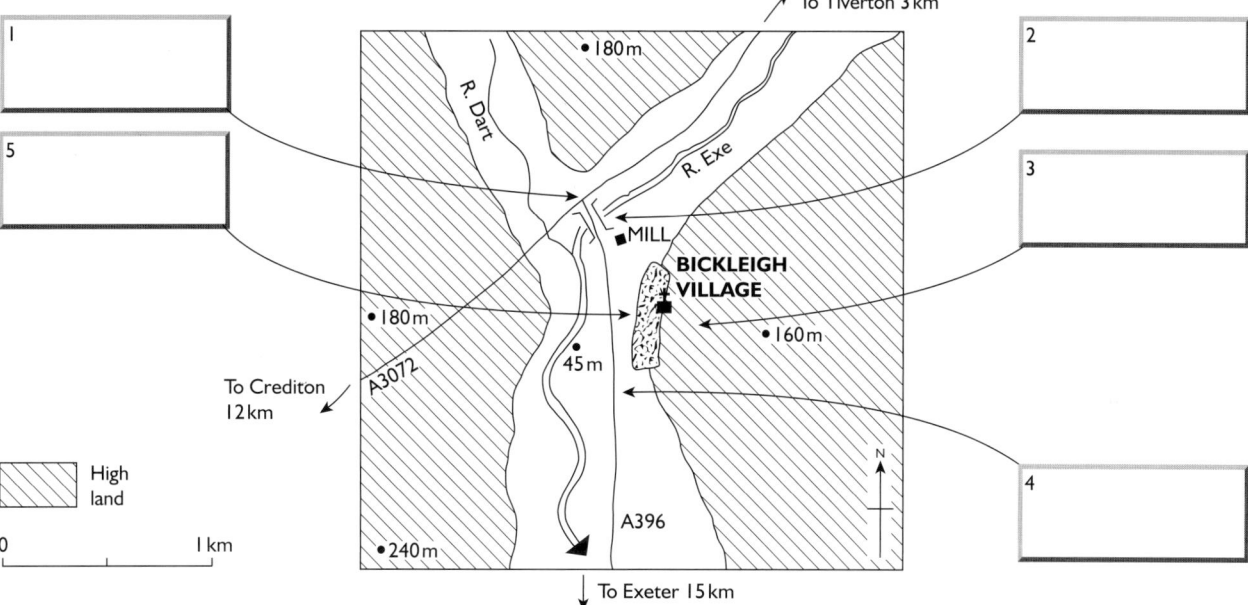

A The site of Bickleigh in Devon.

2 Figure B shows the site of a fishing village in south-west England. Draw a sketch map showing the site of the village. Annotate the map to show five factors which influenced the choice of this site. (5)

3 Draw a sketch map of the village, town or city where you live, showing the main reasons that influenced the original choice of the site. (5)

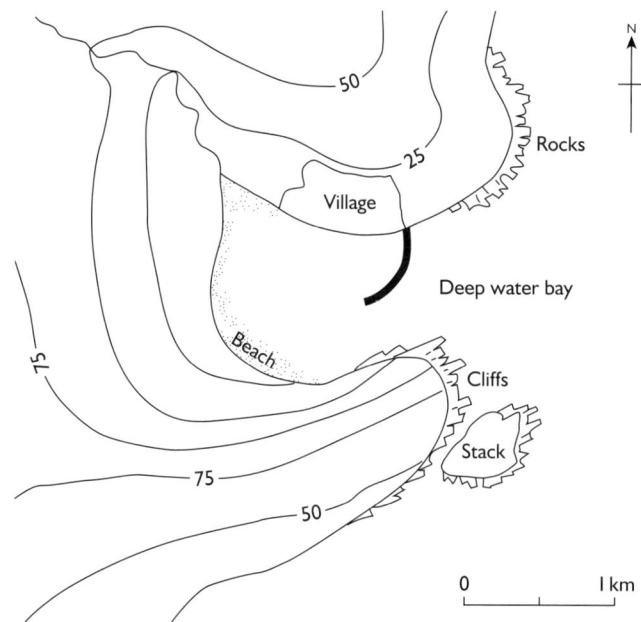

B The site of a fishing village in south-west England.

The Settlement Pattern in Part of East Yorkshire

1 a) Using the OS map of East Yorkshire on page 41 of the pupil book (and the blue figures across the centre), draw sketch maps showing the site and shape of Elstronwick (2232 and 2332); Wyton (1732 and 1733); and Preston (1830 and 1930). (6)

   b) Annotate your sketch maps to show three reasons for the choice of the site of each village. (9)

   c) Which of the three settlements is dispersed, nucleated or linear in pattern? (3)

Functions of Settlement

In the case of the smallest settlements, or hamlets, their function is to simply provide a place for people to live. Larger settlements have additional functions, such as providing shops and services. The number of functions tends to increase with the size of the settlement. Functions also change over time.

Study Figure A, a map of Bridgwater, which is a town of 38 000 people in Somerset.

2 a) One of Bridgwater's most important early functions ceased to exist in the latter part of the 20th century.

   (i) Using evidence from the map, state what you think this function was. (1)

   (ii) Why do you think this function has ceased to exist? (3)

   b) Bridgwater remains an important market town. What advantages does it have for performing this function? (3)

   c) In recent years Bridgwater has experienced industrial growth in the food, drink and distributive sectors. What advantages does the site have for these industries? (3)

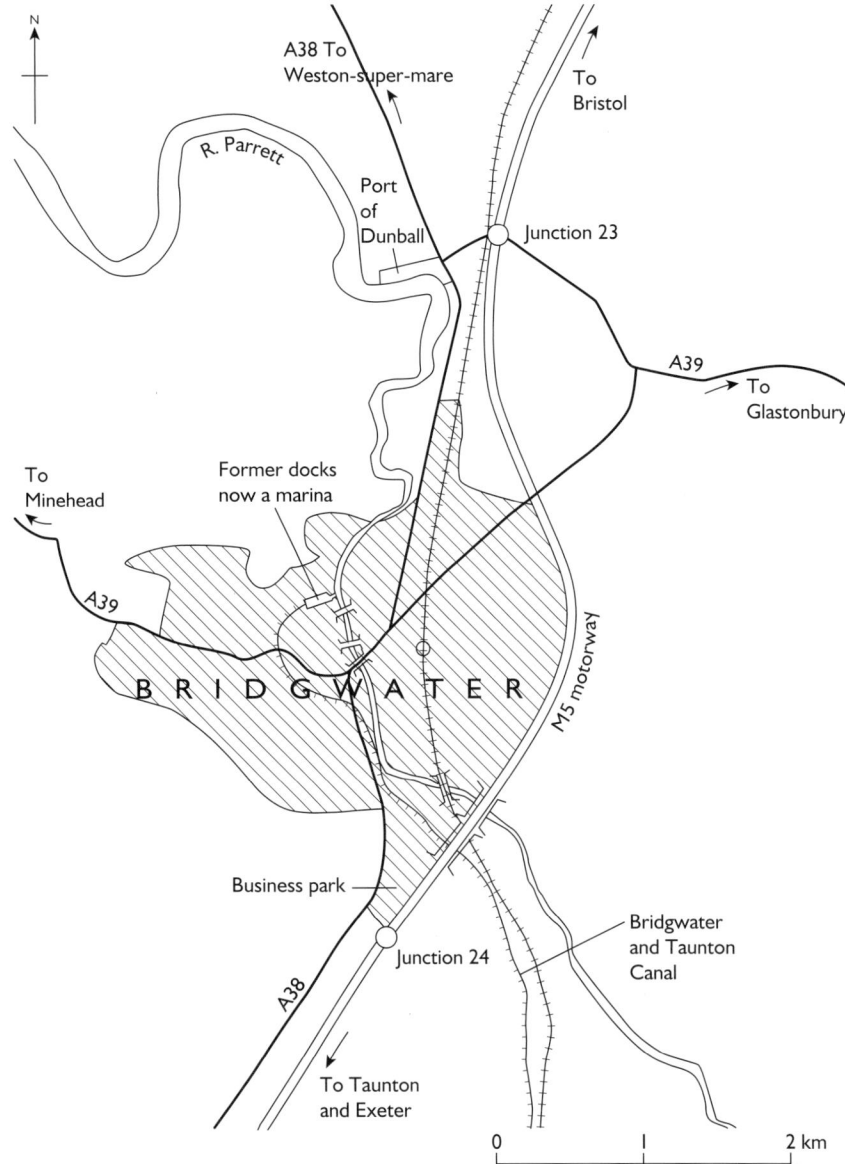

A Bridgwater and surrounding area.

1 Figure A lists six shops and services found in eight neighbouring settlements in a rural area of eastern England.

a) Which settlement is best described as a hamlet? *(1)*

b) Which settlement is a small town? *(1)*

c) What is the population of settlement C? *(1)*

d) How many doctor's surgeries are there in settlement B? *(1)*

e) State two other shops that are likely to be found in settlement F. *(2)*

f) State two other services that are likely to be found in settlement A. *(2)*

Settlement	A	B	C	D	E	F	G	H
Population	8000	5000	1300	1020	850	700	460	380
Shops and services								
Chain store	1	0	0	0	0	0	0	0
Optician	2	1	0	0	0	0	0	0
Doctor's surgery	4	3	1	0	1	0	0	0
Hairdresser	7	5	2	2	1	1	0	0
Post office	3	2	2	1	1	1	1	0
General store	10	7	3	2	2	1	1	1
Total					5	3	2	1

A Shops and services found in eight neighbouring settlements in a rural area of eastern England.

2 a) What is meant by the term *threshold population* for a shop or service? *(2)*

b) What is the threshold of (i) an optician and (ii) a hairdresser in the area shown in Figure A? *(2)*

3 The scattergraph (Figure B) shows the relationship between the population and the number of shops and services found in settlements E to H in Figure A.

a) Complete the totals boxes for Settlements A to D in Figure A. *(4)*

b) Complete the scattergraph by plotting settlements A to D on it. *(4)*

c) Describe and explain the relationship between the number of shops and services and the population of a settlement within this area. *(5)*

4 a) What is meant by the *range* of a service? *(2)*

b) Of the six shops and services listed in Figure A, which two are likely to have the longest range? *(2)*

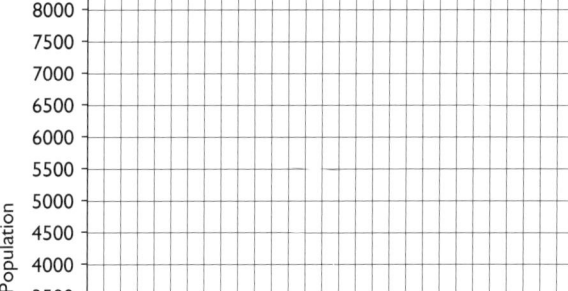
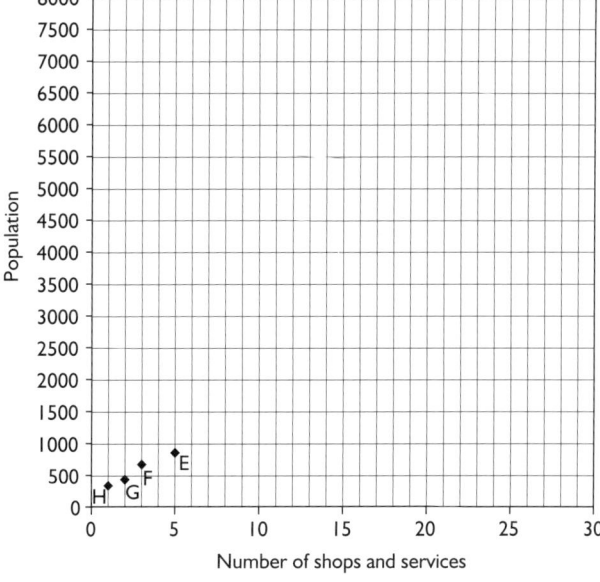

B The relationship between the population and the number of shops and services in the area.

Retailing in the Beverley Area

Beverley is a market town of 25 000 people in East Yorkshire. It has a thriving shopping centre with all the shops and services expected of a town of its size. Nevertheless, it cannot meet all the needs of its inhabitants who sometimes have to travel to neighbouring towns to shop.

1 Study Figure A, a flow-line map showing shopping trips made by people living in Beverley to outside towns.

  a) Approximately what percentage of shoppers travel to (i) Hull, (ii) York and (iii) Willerby?   *(3)*

  b) 10% of trips are made to Hornsea and 5% to Driffield. Add their flow lines to the map, drawn to the correct scale   *(4)*

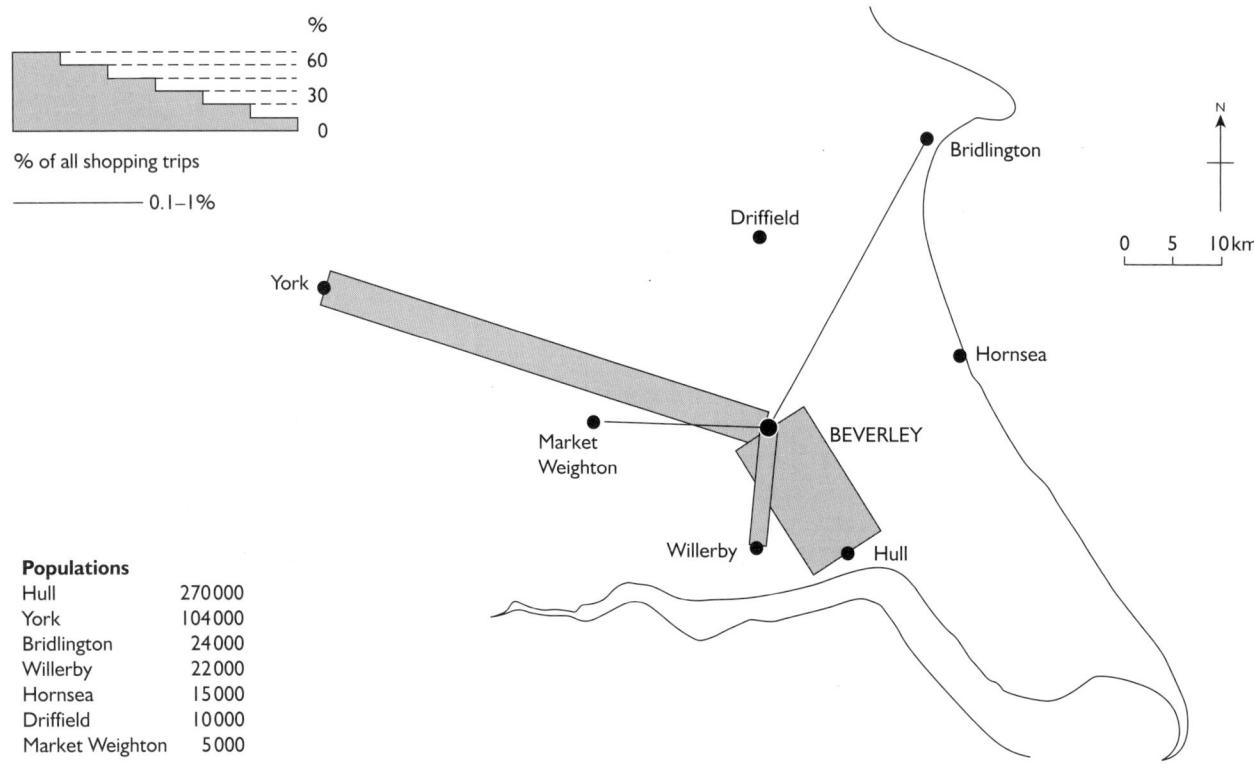

A Shopping trips made by people living in Beverley to outside towns.

2 a) What is the distance from Beverley to (i) Hull and (ii) York?   *(2)*

  b) Why do you think that a fairly large percentage of people travel to York rather than to the closer, and larger, city of Hull?   *(3)*

3 Use the Internet to find information about the shopping centres in York and Hull. Write a paragraph describing the nature and appearance of each. You might start with the following webpages:

www.thisisyork.co.uk/york/marketplace/shopping_in_york.html
www.dooyoo.co.uk/product/129559.html
www.york-tourism.co.uk/visitors
www.hullcc.gov.uk/visithull/shopping.php
www.dooyoo.co.uk/product/68457.html   *(8)*

1 a) What do the letters *CBD* stand for? (1)

b) Complete Figure A by placing the correct urban zone in boxes 1, 2 and 3, choosing from the following:

- Inner City
- Suburbs
- CBD. (3)

c) Why do land values decrease with distance from the city centre? (3)

2 Study Figure B.

a) Name the two most important land uses in each of the three areas. (3)

b) Explain why these land uses are so important in each area. (3)

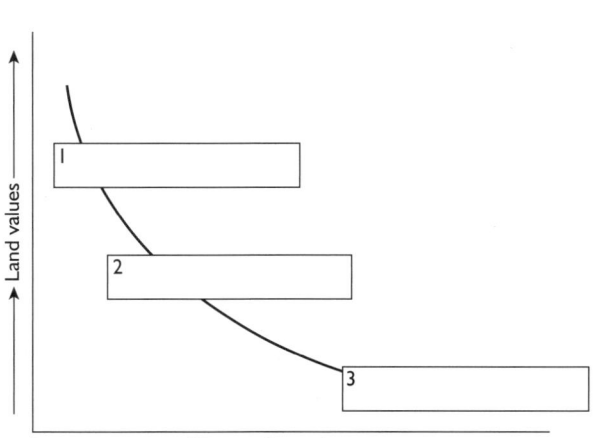

A Land values in a large city.

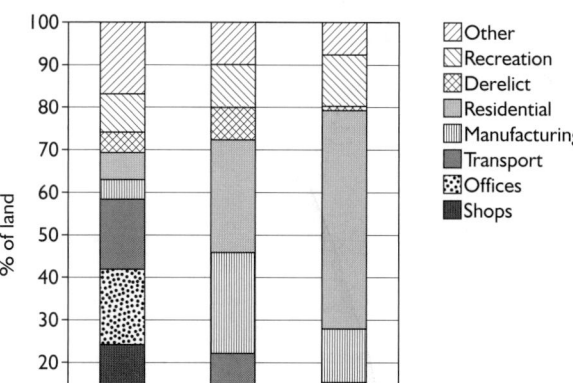

B Land use in a large city.

3 a) Name the person who proposed the Sector Urban Land Use Model shown in Figure C. (1)

b) (i) What is the main factor affecting the shape of this model? (1)

(ii) Explain why this factor creates sectors (or wedges) within the city. (3)

c) (i) Why does modern low-cost housing tend to be located in the same sector as older industrial and low-cost housing? (3)

(ii) Under what circumstances would modern low-cost housing be located in new sectors of the city? Give examples. (5)

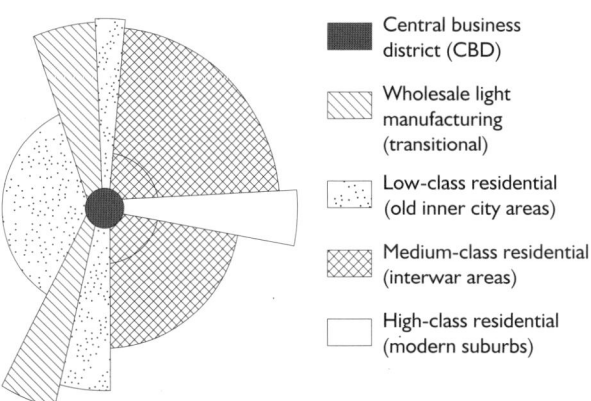

C The Sector Land Use Model.

The Growth of London

New York is the subject of the case study on pages 50–52 of the pupil book. It is a recent city, at least in comparison with its counterparts in Europe. Founded by the Dutch in the 17th century, it was originally named New Amsterdam. The grid-iron pattern of streets, with straight streets meeting at right angles, was the favoured model of town planning in western Europe at the time and the idea was used in New York and became the pattern for all US cities. Figures 3.38 and 3.39 in the pupil book show this pattern very well. Many European cities are much older and their evolution has resulted in a more varied structure.

The Founding of Londinium

London is almost two thousand years old. It dates from A.D. 43 and the Roman invasion of Britain. The Roman soldiers quickly defeated the Britons living in present-day Kent, but to press home their advantage they quickly had to conquer the powerful tribes of Essex and the South Midlands. First, they had to cross the River Thames, but in those days there was no bridge. The Romans followed the river westwards, keeping to the high ground above the marshland bordering the river. Finally the river narrowed and there was a break in the marsh where drier gravelly land reached down to the waterside (see Figure A). At this place the Romans were able to build a bridge and a fort. This became the lowest bridging point of the Thames, that is the last bridge before the open sea. It was also the head of navigation as ships could not pass under the arches of the bridge. As a result this site became an important route centre and seaport. A town rapidly grew, which the Romans named Londinium.

Londinium became the largest city in Roman Britain (with a population of up to 50 000) and it acted as the capital of the province. For a time, it lost that status to Winchester during the later Saxon period, but in 1066 another invader, William the Conqueror, made London capital again. London grew rapidly despite disasters such as the Great Fire of 1666 when much of the old city was destroyed. By 1700, the population of London had reached half a million. As the British Empire grew, London became the greatest seaport in the world. Docks and factories spread eastward along the banks of

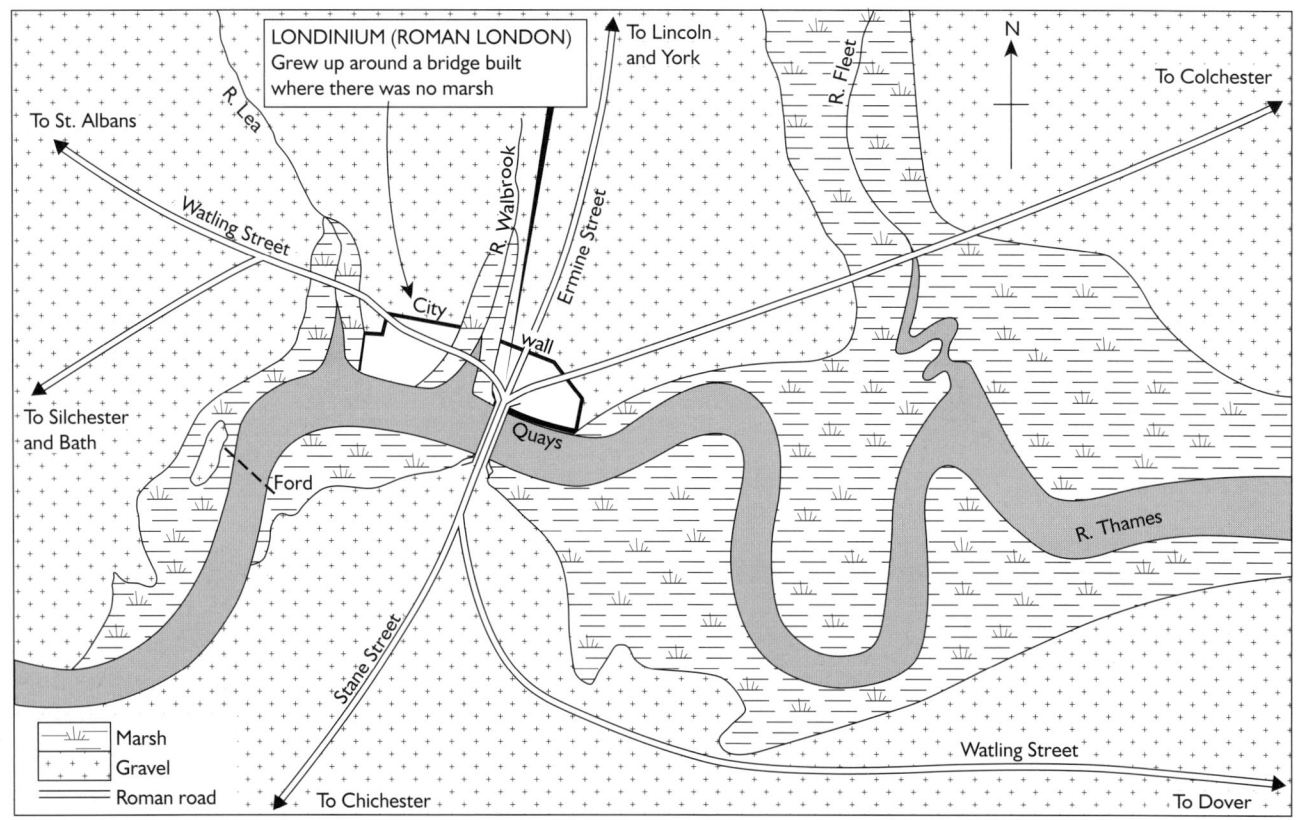

A The site of Roman London (Londinium).

the River Thames as London processed the raw materials of the Empire. At the same time, London became an important commercial, banking and insurance centre.

Urban Growth

London grew especially quickly during the 19th century. New roads and railways linked it with the rest of the country, spreading its influence. In 1888, a new County of London was created to control what had grown to a population of 4.7 million.

The rapid spread of the urban area of the city during the 1920s and 1930s caused great concern – the built-up area tripled in size over two decades (see Figure B). Vast suburbs of houses and gardens sprawled across the countryside surrounding the capital. This growth was permitted by the expansion of suburban underground and surface rail and bus services. The Greater London Plan of 1944 was intended to halt the outward growth of London by:

- the establishment of a green belt of countryside around the city in which building was to be strictly controlled

- a ring of new towns, such as Basildon, Bracknell and Harlow, which were built to take the overspill of population

- directing overspill, from 1952, to expanded towns, such as Swindon, Haverhill and Ashford, beyond the new towns.

In 1978, the focus for London's growth shifted from the area beyond the green belt to the redevelopment of the decaying inner city areas

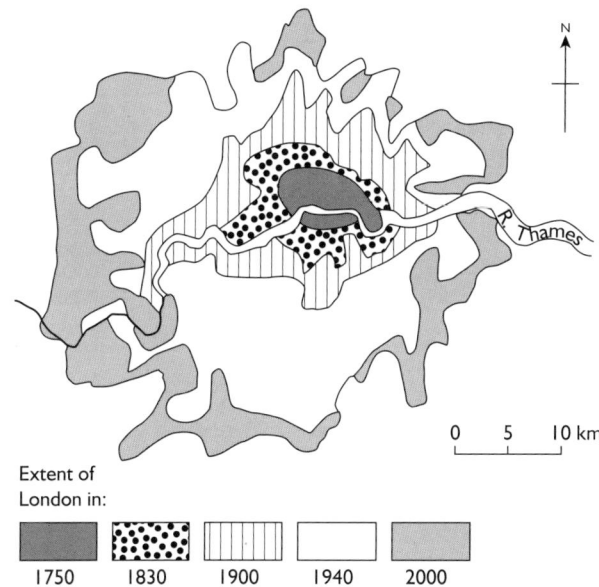

B The growth of London.

Extent of London in:

| 1750 | 1830 | 1900 | 1940 | 2000 |

which had been losing population since the 1950s. London's total population had begun to decline after 1961, when it had reached 8.3 million, and by 1981 it had fallen to 6.5 million. (See pages 60–61 in the pupil book for a description of the regeneration of London's Docklands. The resident population increased from 40 000 in 1981 to 85 000 by 2000.)

Today, London covers over 1500 sq km and has a population of 7.1 million. It is by far the largest city in Britain. The simple structure of London in Figure C shows some similarities to that of Hoyt's Sector Model. This is mainly due to the linear sectors of manufacturing industry.

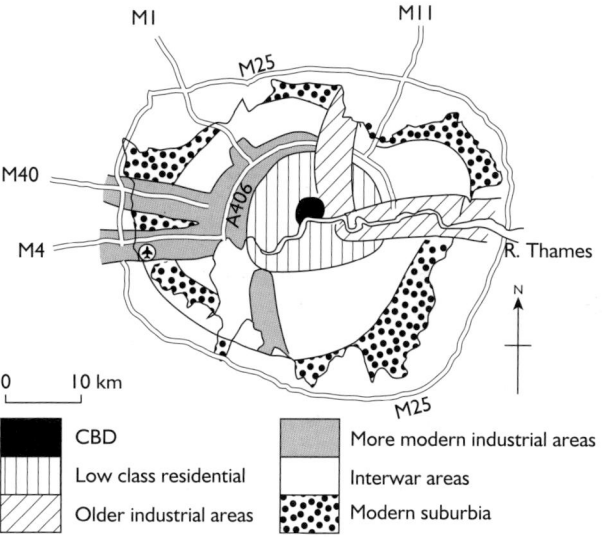

0   10 km

CBD	More modern industrial areas
Low class residential	Interwar areas
Older industrial areas	Modern suburbia

C Simplified urban structure of London.

Industrial Growth

Thames-side industries developed using raw materials imported by ship. They now include flour milling, margarine manufacture, Ford's car engine plant at Dagenham and paper, cement and petrochemical plants. These industries have faced decline in recent years.

The Lea Valley industrial area also developed using the Lea Navigation Canal, which brought bulky raw materials such as coal, timber and steel to the furniture, chemical, metal and engineering works using them. Nowadays good roads provide access and lighter industries have been attracted to the area. Light industries have developed along the main roads radiating out of London. Recently North and West London have developed most because of easier access to Heathrow Airport and to the rest of Britain. The 'Western Corridor' along the M4 is best known for high-tech industries.

Exam Practice Questions

1 Why did the Romans choose the site of Londinium? *(3)*

2 a) What is meant by the terms (i)*the lowest bridging point of a river* and (ii) *the head of navigation of a river?* *(4)*

   b) Why are such sites likely to grow into successful settlements? *(4)*

3 Why did the growth of the British Empire cause London to grow? *(3)*

4 What was the main factor accounting for the rapid growth of London's built-up area during the 1920s and 1930s? *(1)*

5 How did the Greater London Plan of 1944 propose to handle London's future growth? *(3)*

6 a) Using the statistics in Figure D, draw a line graph showing the size of London's population between 1801 and 2001. *(5)*

   b) Describe and explain the changes in population shown in your graph. *(5)*

Year	1801	1811	1821	1831	1841	1851	1861	1871	1881	1891	1901	1911
Population (millions)	1.1	1.3	1.6	1.9	2.2	2.7	3.2	3.8	4.7	5.6	6.5	7.1

Year	1921	1931	1939 (est.)	1951	1961	1971	1981	1991	2001(est.)
Population (millions)	7.4	8.1	8.6	8.3	8.2	8.1	6.7	6.7	7.2

D The population of London, 1801–2001.

7 Study Figure E, which shows how the population density for a major world city in Europe has varied over a period of 120 years. The lines for each year are called population density gradients.

   a) Why did the population density in the inner city decline between 1930 and 1980? *(4)*

   b) Why did the density in the outer city rise between 1930 and 2000? *(4)*

   c) Why did the population density in the inner city increase between 1980 and 2000? *(4)*

8 For a named major city in the EU, describe the planning measures adopted to control and redirect urban growth. *(7)*

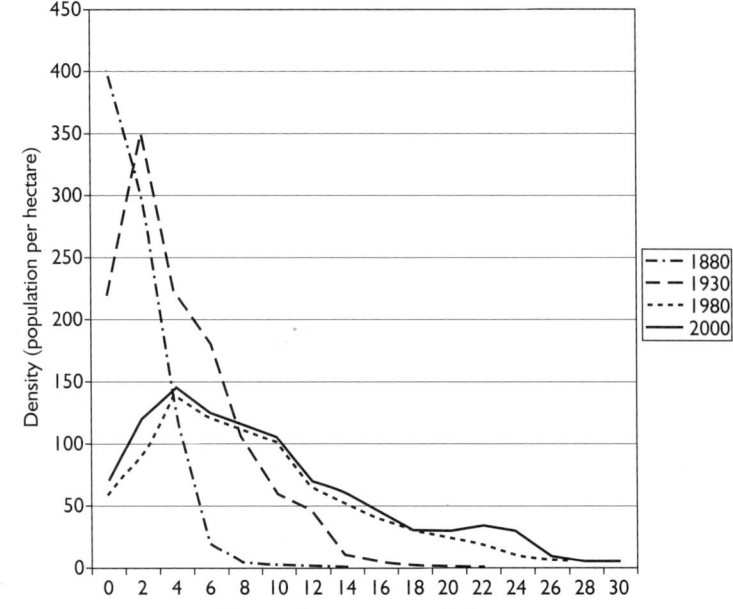

E How the population density for a major world city in Europe has varied over a period of 120 years.

The Growth of London

These model answers are designed to show you how to get full marks for the longer questions in the Case Study Extra for this section. Read the question carefully and write your own answer. Then read the model answer and the examiner's notes to see what he or she is looking for when awarding the marks. Decide how many marks you think the examiner would give your answer. Decide how to change your answer to this particular question to increase your marks. What will you do differently next time you answer a similar question?

7 Study Figure E, which shows how the population density for a major world city in Europe has varied over a period of 120 years. The lines for each year are called population density gradients.

a) Why did the population density in the inner city decline between 1930 and 1980? *(4)*

> In 1930 much of the inner city still consisted of congested terraced housing, some of which had become slums. As people's incomes rose they were able to consider moving out of the inner city to more attractive, modern housing in the suburbs. Councils undertook redevelopment schemes which involved the demolition and clearance of large areas of 19th century slums and their replacement with lower density housing.

Level 1 (1–2 marks): *Refers to slum clearance.*
Level 2 (3–4 marks): *Additionally refers to people choosing to move to more modern, attractive housing in the suburbs.*

b) Why did the density in the outer city rise between 1930 and 2000? *(4)*

> The process of suburbanisation had already begun as people were able to travel longer distances to work because of the expansion of suburban rail and bus services. Urban sprawl was the result. After the Second World War the increase in car ownership and improved road systems allowed people to travel even further to work. More modern housing with gardens and all modern conveniences was built at a lower density in the suburbs. In addition to voluntary movements, many of the inhabitants displaced by slum clearance schemes were moved to council estates built near the city boundary.

Level 2 (1–2 marks): *Refers to suburbanisation.*
Level 3 (3–4 marks): *Explanation of suburbanisation including outer city council estates.*

c) Why did the population density in the inner city increase between 1980 and 2000? *(4)*

The policy of slum clearance adopted in the 1960s and early 1970s had attracted much criticism because of the break up of communities and the poor quality of many of the high-rise flats. Councils started to follow urban renewal schemes by improving existing properties. Private developers did the same. During the 1980s, it became fashionable for young middle class couples to live in new housing built in the inner city, e.g. London Docklands.

> **Improve Your Mark!**
>
> **Level 1 (1–2 marks)**: *Reference to urban renewal.*
> **Level 2 (3 marks)**: *More detail on urban renewal.*
> **Level 3 (4 marks)**: *Additionally includes criticism of slum clearance leading to change of policy.*

8 For a named major city in the EU, describe the planning measures adopted to control and redirect urban growth. *(7)*

The built-up area of London tripled in size between 1920 and 1940. In order to control and re-direct this growth, the Greater London Plan of 1944 established a 'green belt' around the fringe of the city. Within the green belt, new building was strictly controlled in order to preserve the countryside and halt the urban sprawl.

It was recognised that there would still be great demand for new housing and industry around London, so a ring of new towns such as Basildon and Bracknell were built beyond the green belt to take the overspill of population. From 1952, overspill was directed further out to existing towns such a Swindon and Ashford, which were called expanded towns.

From 1978, growth was encouraged in the decaying inner city areas. Huge redevelopment schemes took place, notably in London's Docklands. Planning controls were relaxed in the area of the Isle of Dogs, which was designated as an enterprise zone. The London Docklands Development Corporation was established to regenerate the area for commerce and housing. The resident population increased from 40 000 in 1981 to 85 000 by 2000.

> **Improve Your Mark!**
>
> **Level 1 (1–2 marks)**: *Reference to planning controls, including green belts, designed to stop growth.*
> **Level 2 (3–5 marks)**: *Reference to a valid named city with an awareness of the need to channel growth elsewhere.*
> **Level 3 (6–7 marks)**: *A fuller answer including more policies, plus some supporting examples and/or statistics.*

CHAPTER 4 deals with urban change in developed countries. This is covered under four sections: changes in the CBD, in old inner city areas and at the rural-urban fringe, and transport in urban areas. The final section investigates Osaka-Kobe as a case study of urban change. Other studies in the chapter include the regeneration of London's Docklands, the Gateshead MetroCentre as an example of an out-of-town shopping centre, Thurston in Suffolk as a suburbanised village and the integrated traffic system of Lille in France.

Changes in the CBD   58 – 59

The focus here is on recent changes. A useful exercise is to compare the pupils' local CBD with the examples shown in Figures 4.1 to 4.5 on pages 58–59. Encourage active discussion on what features they have in common.

Activity Sheet 4.1 enables pupils to look in detail at service changes in the CBD. It provides a sketch map of part of the High Street of a town in Norfolk in 1980 and 2003, intended to clarify the changes occurring in a typical British CBD. Significant changes which could be mentioned in answer to Question 2 include the pedestrianisation of the street, with seats and a fountain intended to make shopping a safer and more pleasant leisure activity rather than an everyday chore; the replacement of small, probably individually-owned shops by a national chain; the disappearance of the butcher, baker and grocer and their replacement with a small supermarket, showing the advantages of economies of scale; the disappearance of the café and its replacement by a fast food outlet, reflecting changing tastes and needs; the appearance of computer, video and mobile phone shops, reflecting the development of these technologies since 1980.

Changes in Old Inner City Areas   60 – 61

Following a brief introduction describing the establishment of urban development corporations, there is a detailed study of the regeneration of London's Docklands. Once the background is established, progress since 1981 is covered in terms of physical, environmental, economic and social regeneration. A list of groups helping in the regeneration of the Docklands is followed by a consideration of the conflicting opinions of the results. There is potential to use role play here with the following characters: a young local person, an elderly local, 'yuppie' newcomers, property developers and a former docker.

Activity Sheet 4.2 covers a simple classification of urban development programmes and supports the London Docklands study.

Changes at the Rural-Urban Fringe   62 – 63

The focus here is on development of greenfield sites at the edge of urban areas. There is plenty of opportunity to use examples from the pupils' local area and role plays to support this study.

Activity Sheet 4.3 provides some useful statistics to back up the changing nature of retail sales by location in the UK. Figure 4.11 on page 62 shows a retail park beside a motorway junction in Carlisle. Activity Sheet 4.3 builds on the same theme. Retail parks such as Hankridge Farm are more typical of most British town and city centres than the grand scale shopping malls such as the MetroCentre, Meadow Hall or Blue Water Park are. The sheet demands a range of skills including use of the Internet to research the opinions of a pressure group.

Suburbanised villages   64 – 65

On this page the focus is on suburbanised villages. Figure 4.15 on page 64 is a very useful table summarising the changes affecting such villages. A short case study of the village of Thurston in Suffolk builds on this. Figure 4.16 on page 64 gives maps that clearly show the importance of farmland sales to the growth of the village between 1884 and 1994. The changing nature of the village inhabitants is crucial – few now work locally and most commute to work in Bury St. Edmunds or Stowmarket, or further afield to Ipswich or Cambridge. A useful additional activity is for the pupils to use an atlas to locate Thurston (5 km east of Bury St. Edmunds) and suggest where the new villagers commute to work.

Activity Sheet 4.4 uses a newspaper article to elicit information on counter-urbanisation and its effects. The article has a high language level and the first question probes the pupils' understanding of some of the complex words and phrases used, so this needs to be carefully used with pupils of less than high ability. The activity makes further use of the case study of Thurston from the pupil book in order to draw out the major points about the changing function of the village.

Brownfield and Greenfield Sites   65

This page acts as a useful summary of the issues arising from urban sprawl, urban regeneration and counter-urbanisation. By focusing on the shortage of housing in Britain, the page neatly encapsulates the dilemma between building on brownfield or greenfield sites. Clearly, there will be opportunities to use local examples. Estate agents, advertisements for new housing developments can be a very useful resource here.

Pages 66 and 67 focus on commuting. Note the recent trend for 'reversed' commuter flow (page 66) – pupils referring accurately to this in examination answers could gain additional credit. Many pupils will have their own experiences of commuting to draw from. Teachers should use this to build up a list of reasons for commuting and the resulting issues, then refer the pupils to Figure 4.23 on page 67.

Figure 4.24 is a good starting point for the discussion of solutions. Can the pupils see any problems arising from the government's Ten-Year Transport Plan?

Pages 68 and 69 use the case study of Lille as an example of an integrated traffic system. The influence of one person's vision in persuading French railways to build their new high speed train station in the centre of the city, rather than on the outskirts, can be highlighted. The VAL underground metro system is a modern rapid transit system for the inner urban area, which means that people do not need to use their cars. The comprehensive nature of the Eurolille development at the station site should be emphasised. Transport is an integral part of Lille's inner urban area.

Activity Sheet 4.5 requires work with a suitable spreadsheet package to produce graphs, although this could be done on paper if necessary. The first activity concerns the mode of transport used for journeys to work into English cities. The predominance of road transport, and especially of the car, is clear. The second activity highlights the importance of the rush hour in urban traffic flows. There are opportunities for fieldwork here, with traffic counts on urban roads at different times of day, to produce graphs similar to that included with the sheet.

Case Study 4 Osaka-Kobe 2000   📖 70 – 73

Osaka-Kobe serves as a case study of the problems facing major urban areas and of some effective solutions, which include building on land reclaimed from the sea. The Shinkansen high speed trains provide efficient inter-city transport, as does Kansai International Airport, built on reclaimed land. Within the conurbation, new roads and bridges have been built.

Case Study Extra Urban Change in Paris

Paris provides a contrasting example to Osaka-Kobe. The Japanese conurbation is composed almost entirely of 20th and 21st century buildings; Paris has a long historical legacy of architecture and transport arteries. The case study of Paris shows the extent of the problems facing Paris and the imaginative schemes, at a variety of scales, which have been adopted to overcome them.

1 a) What do the letters *CBD* represent? (1)

  b) Name two features of CBDs that are not usually found in other parts of the inner city. (2)

2 Figure A shows the high street in the CBD of a Norfolk town in 1980 and in 2003.

  a) List five changes between 1980 and 2003. (5)

  b) Explain why these changes have taken place. (10)

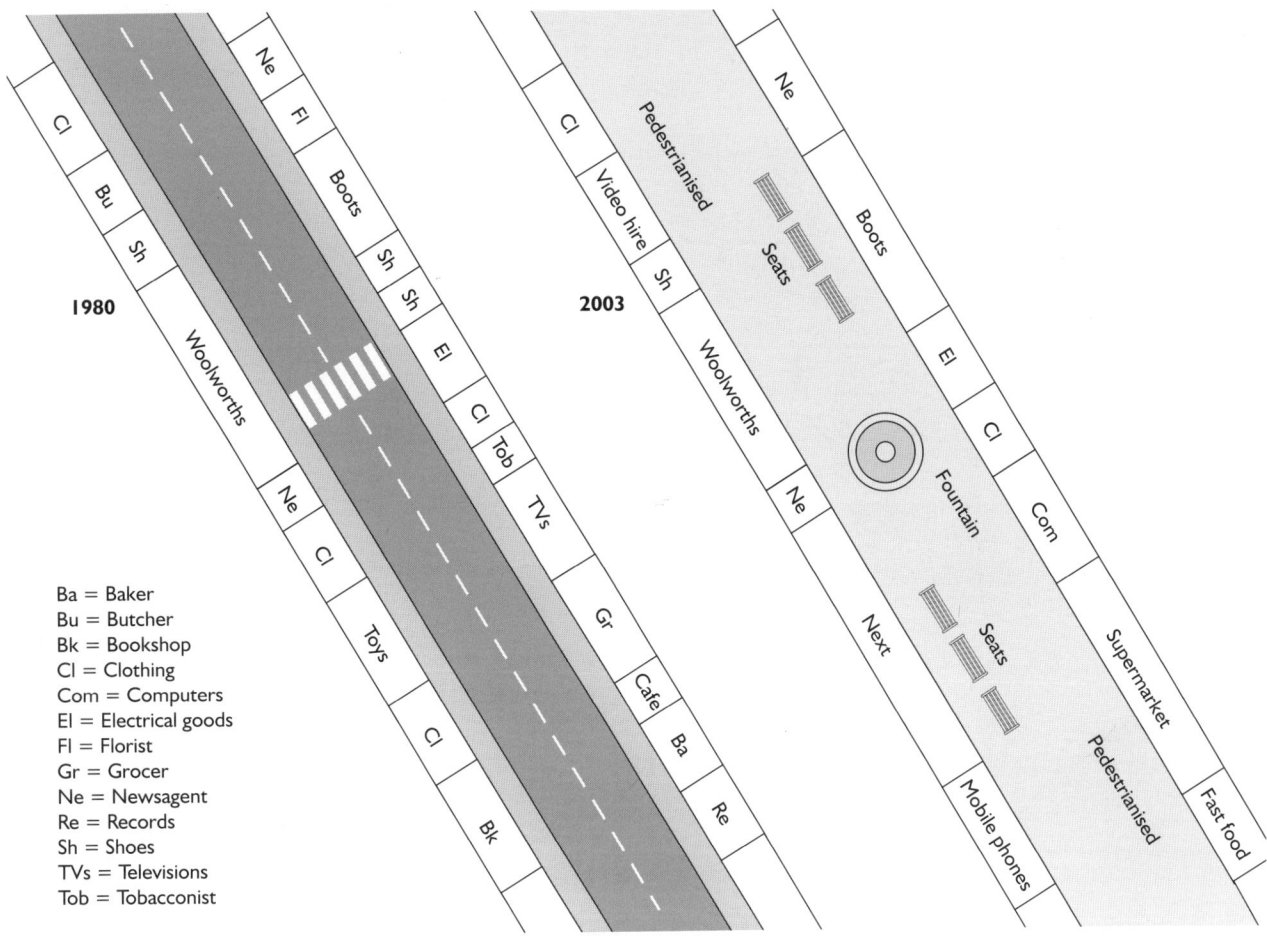

Ba = Baker
Bu = Butcher
Bk = Bookshop
Cl = Clothing
Com = Computers
El = Electrical goods
Fl = Florist
Gr = Grocer
Ne = Newsagent
Re = Records
Sh = Shoes
TVs = Televisions
Tob = Tobacconist

A Part of the high street of a town in Norfolk in 1980 and 2003.

3 Why do certain types of land use, such as shoe shops, tend to be clustered within the CBD, whereas others, such as newsagents, are more widely distributed? (2)

1 Figure A shows a simple classification of urban development programmes.

   a) Describe two problems that can be caused by the expansion of towns and cities out into the countryside.   (4)

   b) Comment on one planning policy (such as green belts, urban renewal, etc.) used to control urban sprawl, giving an example of a real development to support your comments.   (4)

   c) Give an example of each of the four developments in the lowest section of Figure A.   (4)

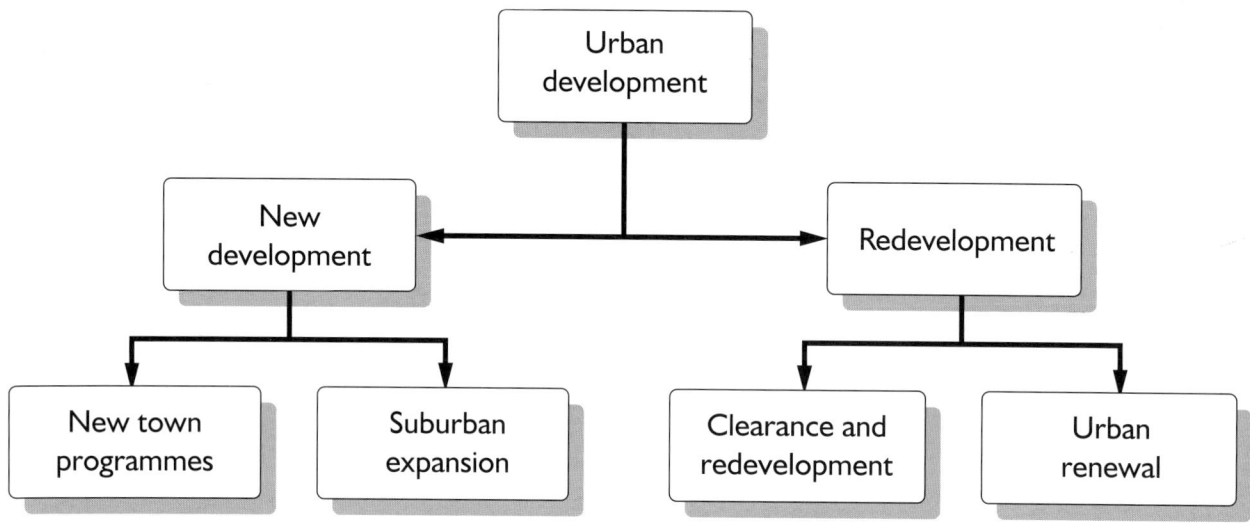

A A simple classification of urban development programmes.

2 Describe two of the aims of urban clearance and redevelopment programmes.   (4)

3 Why is urban renewal now preferred to clearance and redevelopment programmes in many cities?   (3)

4 For one large urban redevelopment programme you have studied:

   a) describe the conditions which caused the programme to be undertaken and the main actions taken   (6)

   b) list some of the successes and failures of the programme in a table with two columns, one headed 'Good Points' and the other headed 'Bad Points'.   (6)

1 Complete Figure A, which is a graph of retail sales in the UK by location using the following data. *(6)*

Year	1996	1997	1998	1999	2000
High Street	47	46	45	45	44
Neighbourhood	26	25	24	23	22
Out-of-town	27	29	31	32	34

2 What does the graph show about the changes in retail sales in the three types of shopping centre? *(3)*

3 Give three reasons that help to explain these changes. *(3)*

Figure B shows the out-of-town shopping centre at Taunton in Somerset. Called Hankridge Farm, it provides an interesting contrast to the very large, regional out-of-town centres such as the MetroCentre covered on page 63 of the pupil book. Although less dramatic in scale than the MetroCentre, it is more typical of the retail parks found on the rural-urban periphery of many British towns and cities. It contains a large 'superstore' belonging to one of the leading supermarket companies. Some people and organisations object to such developments; visit the Council for the Protection of Rural England's website at www.cpre.org.uk/policy/positions/majretail.htm to research some opinions of this nature.

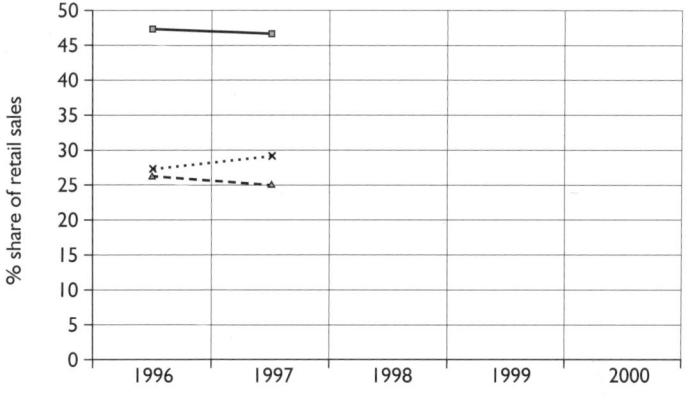

A Retail sales by location 1996–2000.

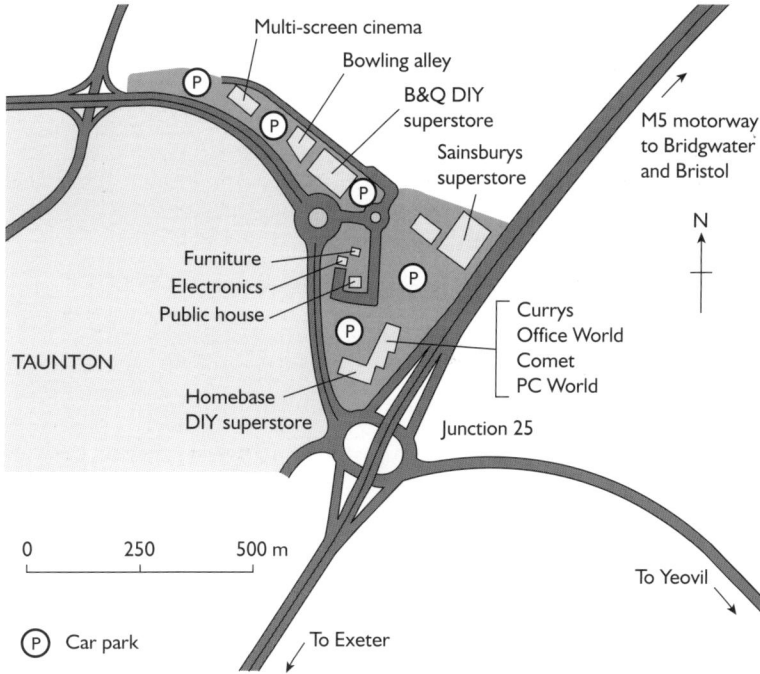

B Hankridge Farm Retail Park, Taunton, Somerset.

4 a) Hankridge Farm was built on former farmland in the flood plain of the River Tone. What name is given to such a development site? *(1)*

b) What advantages does this site have for an out-of-town shopping centre? *(4)*

c) Why do some people object to the use of such sites for the development of out-of-town shopping centres and retail parks? *(5)*

Read this newspaper article carefully. Although written several years ago, it is still directly relevant to the rural/urban issues of today.

The cosy image of the English countryside hides increasing problems of rural poverty and homelessness, campaigners warned yesterday. Influxes of the rich and retired in search of a pastoral dream are threatening attempts to improve the quality of life for hard-up families and young people, they said.

There is a conflict over the countryside between those who want it as a nice rural museum, in which they can be retired, go for holidays or commute to jobs somewhere else – and the indigenous rural population who used to work on the land and now need other jobs. That battle is being lost at tremendous speed. Les Roberts, director of the pressure group ACRE (Action with Communities in Rural England) said: 'in seemingly gentrified villages with BMWs on every drive, there is also a population on low income, struggling to survive, and yet half the people there will say "there is no problem, we go to Sainsbury's in the car and it's a lovely place to live". Such an attitude helps to prevent the countryside getting its fair share of government aid for regeneration.'

From **The Guardian** *13 July 1995.*

1 What do you understand by the following phrases:
   - 'influxes of the rich and retired'
   - 'in search of a pastoral dream'
   - 'the indigenous rural population'
   - 'gentrified villages'.   *(4)*

2 a) What does the term *counter-urbanisation* mean?   *(2)*
   b) Suggest two reasons why counter-urbanisation is occurring.   *(4)*
   c) What types of people are most likely to be 'counter-urbanites'?   *(2)*

3 a) Give three disadvantages of rural areas resulting from counter-urbanisation. *(3)*
   b) Give three advantages of rural areas resulting from counter-urbanisation.   *(3)*

4 Refer to the text and maps on Thurston on page 64 of the pupil book.
   a) Why did Thurston begin to grow rapidly from the 1950s?   *(1)*
   b) What transport development encouraged further growth?   *(1)*
   c) How have the functions of Thurston changed since 1884?   *(4)*

Most world cities suffer from transport problems. In many cases, the growth in numbers of road vehicles has brought traffic to a near standstill. Governments everywhere seek solutions. The integrated transport system of Lille is studied on pages 68–69 of the pupil book.

1 Study Figure A, which shows the modes of transport used for journeys to work into cities in England.

a) Enter the data into a spreadsheet program and produce (i) a bar graph and (ii) a pie graph. Print both graphs. Which do you think is more effective? Why? *(7)*

b) What percentage of journeys is by public transport? *(1)*

c) What percentage is by car? *(1)*

d) List four problems that may be caused in urban areas by the high percentage of journeys by car. *(4)*

Mode	% of total
Train	4.2
Underground railway	2.2
Bus	9.7
Car driver	56.5
Car passenger	7.7
Motorcycle	1.6
Pedal cycle	3.3
Foot	12.2
Other	2.6

A Modes of transport used for journeys to work into cities in England.

2 Figure B shows the traffic flow into Leeds on a typical weekday.

a) Describe the pattern shown by the graph. *(4)*

b) Why is there a steep rise in the volume of traffic after 7 a.m.? *(1)*

c) (i) At what time is the peak of the morning rush hour? *(1)*

(ii) How many vehicles per minute are entering the city at that time? *(1)*

(iii) Compare the shape of the graph and the number of vehicles at the morning and evening rush hours. Why do they appear to be different? *(4)*

3 Use the Internet to discover the basic facts on the congestion tax introduced in London in February 2003. Design a poster promoting the benefits of the scheme. *(4)*

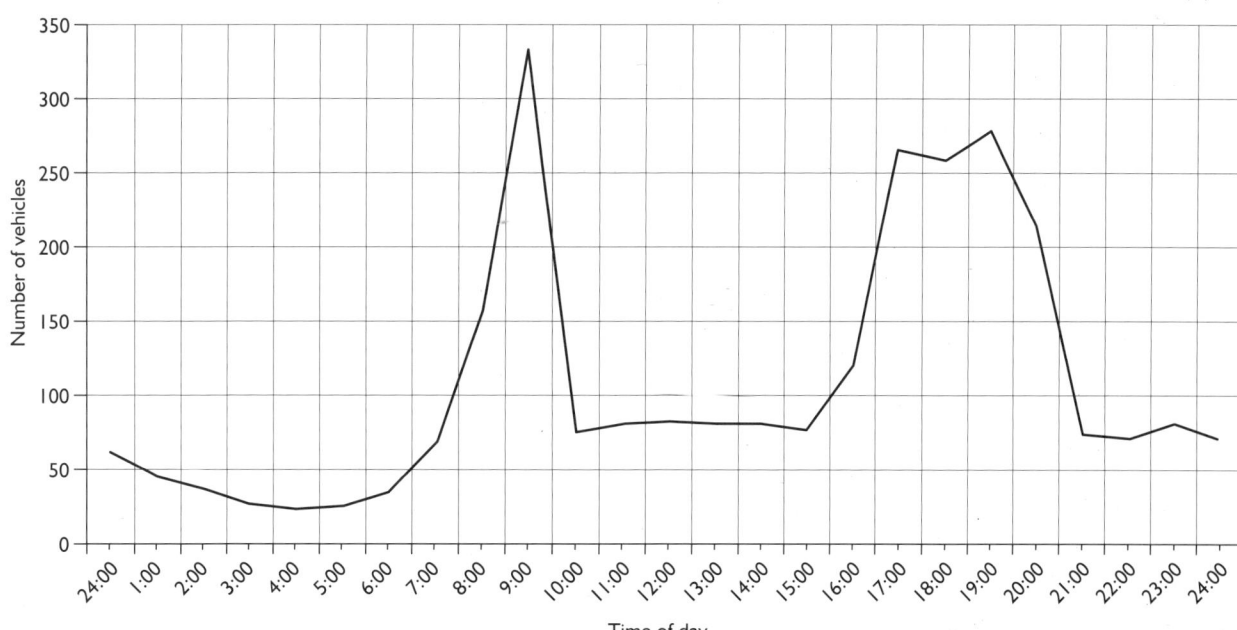

B Traffic flow into Leeds.

Urban Change in Paris

Paris provides a contrasting case study to that of Osaka-Kobe on pages 70–73 in the pupil book. The Japanese conurbation is composed almost entirely of 20th and 21st century buildings; by contrast, Paris has much older surviving buildings and transport arteries.

With its population of 10.5 million in Greater Paris, this is a city of world importance. Paris faces the problems of poor housing, high land values, congested transport systems and pollution similar to any major city. The solutions adopted have been extensive and innovative.

Housing

Paris grew rapidly during the 19th and early 20th centuries. The city became overcrowded and congested. During the late 19th century, row upon row of tiny terraced houses and tenement apartments were built to house the growing industrial workforce. These buildings were cramped and often lacked all basic services.

Between the two world wars, houses and apartments spread outwards from the inner industrial suburbs and within those two decades the built-up area of Paris doubled. These houses were often better built than the earlier slums, but the developments were unplanned, lacking shops, hospitals, schools and even roads.

After the Second World War, huge estates of high-rise apartment blocks called grands ensembles were built. The largest is Sarcelles, which houses 80 000 people. The building of services here was slow. The roads and railway lines that surround the grands ensembles usually isolate them from the rest of the city. There were severe social problems caused by living in the environment of these concrete tower blocks.

In 1961, it was estimated that half the homes in Paris consisted of two rooms or less. Over half lacked running water and toilet facilities. Central Paris had a population density of 35 000 people per sq km (in the same year the density in central London was 10 000 people per sq km).

Transport infrastructure

By 1970, the transport infrastructure in Paris faced several problems.

■ The roads, apart from the famous broad boulevards such as the Champs-Elysées, were narrow and inadequate for modern road traffic. There were many complex junctions where lots of roads converged at a single point.

■ In a region with one of the highest rates of car ownership in Europe, there were not enough car parking spaces. This forced people to park on the street and on the pavements.

■ The mainline railway stations clustered around inner Paris, with their entrances on small offshoots of the boulevards. As a result they created severe traffic jams.

■ The Paris underground railway, the Metro, was not connected to the main railway system and did not extend far outside central Paris.

Taken together, these factors resulted in gridlock.

The solutions

Unlike London with its constricting green belt, Paris' planners allowed the city to continue to grow. The growth was to be directed into two corridors (see Figure A). Five new towns (shown on Figure A) were then allowed to develop within the corridors of growth. The new towns now have a full range of shops and services, and employment is provided in offices and business parks which are carefully zoned away from housing areas. Elsewhere, new development was discouraged.

After nearly twenty years of trying to hold to this plan, the policy was changed during the 1980s and growth was encouraged within Paris again, in an attempt to compete with rival European conurbations, especially London.

A Plan of Paris.

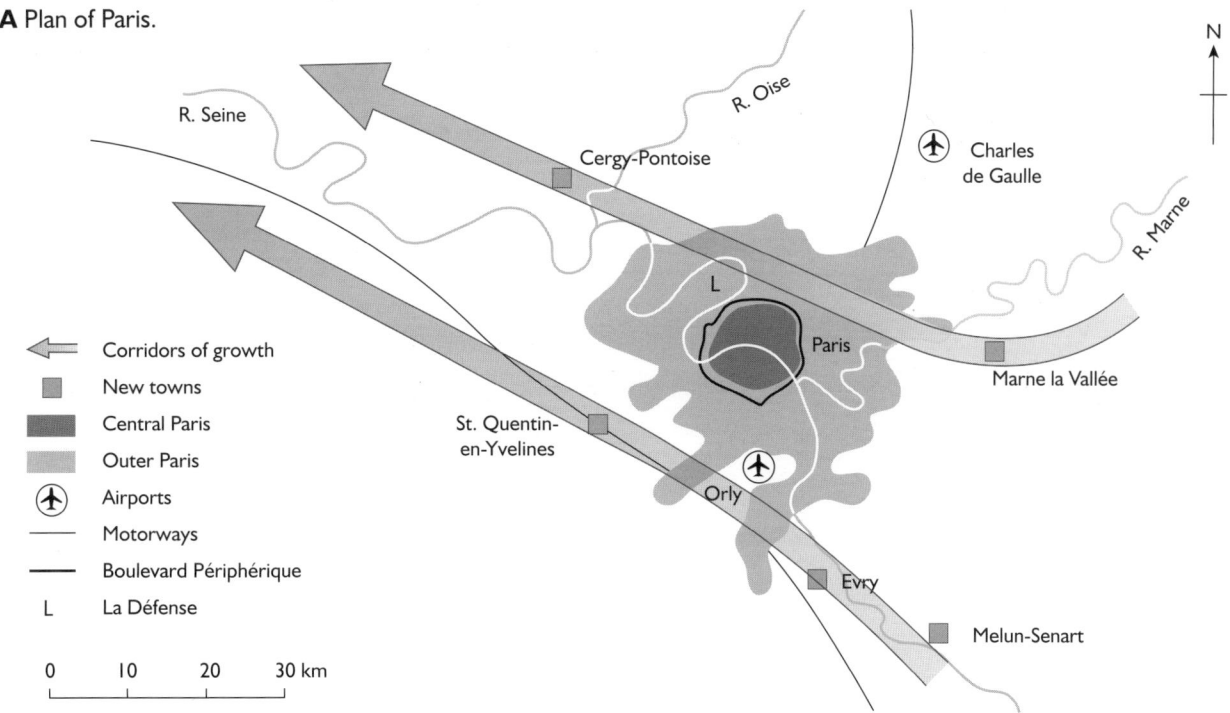

Within Paris, much urban redevelopment took place. The worst slums were demolished and cleared. Urban renewal in other areas included the building of services, better public transport and more public open spaces. Several large-scale developments took place including:

- La Defense, a complex of 25 skyscrapers containing offices, shops and apartments. This has become the largest office development in Europe. The architecture is dramatic, including the shell-like Palais which houses a centre for industry and La Grande Arche, an enormous hollow cube housing a conference centre and exhibition gallery. 50 000 people work here and 20 000 live in the apartments.

- The fruit and vegetable markets at Les Halles were moved out to a new site in the suburbs at Rungis. The old market site has been transformed into a five-level complex of shops, apartments, gardens and leisure facilities. The modernistic Forum des Halles attracts the young, seeking the latest fashions within the concrete and glass bubbles of the underground arcades. This is the most intensively developed retail area, generating more money per square metre than any other shopping centre in France. There are also six hundred new apartment homes in the complex.

- Near Les Halles, in the Beauborg district, stands the Pompidou Centre, opened in 1977 and completely renovated for 2000. This major arts centre revitalised the formerly rundown area. The smaller streets around the Centre are full of art galleries and fashionable shops and cafés.

- At Maine-Montparnasse a 210 m high black skyscraper is the centre of another development of shops and offices, providing over 40 000 jobs.

- At Bercy developments include the Palais Omnisports stadium, new residential developments and a large new park opened in 1997, only the second large park to be opened in Paris since the 1860s.

Transport improvements have included the completion of an orbital motorway, the Boulevard Peripherique, the improvement and modernisation of the existing Metro system and the construction of a new high speed Metro, the RER. Within Paris, more car parking spaces have been provided. The French Government and the Parisian local authorities both subsidise public transport.

The RER serves the outlying suburbs, the airports and even EuroDisney. It also connects with the old metro at certain stations. In 1998 the first 7 km of Paris' new Meteor Metro line, Line 14, was opened. It was soon carrying 130 000 passengers per day between south-west and north-east Paris. The Meteor provides some much needed relief for the Metro Line 1 and the RER Line A which run crowded all day long.

The Meteor's first section cost £800 million; it is a fully automated system with no drivers. Ultimately the Meteor will run out to Paris (Orly) airport.

Exam Practice Questions

1 a) Construct a line graph to illustrate the statistics in Figure B. *(5)*

b) What housing problems were created by the rapid growth of Paris between 1880 (population 2 million) and 1962? *(6)*

c) What possible factors may explain the city's slower growth since 1971? *(3)*

Year	1954	1962	1971	1985	2002
Population (millions)	6.7	7.8	9.5	10.1	10.5

B The population of Paris, 1954–2002.

2 Describe the transport problems that Paris faced by 1970. *(4)*

3 What have been the main features of the urban redevelopment projects within Paris? *(5)*

4 Describe the improvements made to Paris' transport infrastructure since 1970. *(6)*

Urban Change in Paris

These model answers are designed to show you how to get full marks for the longer questions in the Case Study Extra for this section. Read the question carefully and write your own answer. Then read the model answer and the examiner's notes to see what he or she is looking for when awarding the marks. Decide how many marks you think the examiner would give your answer. Decide how to change your answer to this particular question to increase your marks. What will you do differently next time you answer a similar question?

1 b) What housing problems were created by the rapid growth of Paris between 1880 (population 2 million) and 1962? *(6)*

> Housing: Poor quality housing with few if any basic services were rapidly built to house the population growth during the late 19th and early 20th centuries. This housing rapidly became slums. Between the world wars, better built but unplanned housing developments sprawled across the surrounding countryside, lacking shops and services. From the 1950s, huge estates of high-rise apartments called grands ensembles, such as Sarcelles, were built; they have the well-known social problems caused by living in cheap, high-rise developments. In 1961, over half of the homes in Paris lacked private running water and toilet facilities.
>
> Pollution: Pollution from increasing traffic fumes and from factories and workshops built within the residential areas made Paris' atmosphere dangerously unhealthy.

Level 1 (1–2 marks): *Reference to slums and/or road congestion and pollution.*

Level 2 (3–4 marks): *Greater detail on the housing issues, with some attempt to differentiate between periods and types of housing development.*

Level 3 (5–6 marks): *Detailed and analytical answer including the specific problems associated with the different periods and types of housing development, and the nature of the traffic congestion problems.*

3 What have been the main features of the urban redevelopment projects within Paris? *(5)*

> The worst slums were cleared; others were improved, with new services and more public open spaces. A number of large-scale projects have been undertaken, usually multi-functional schemes including shops, services, residential developments and office employment; most include dramatic modern architecture. Transport improvements have been integrated within many of these projects. An example is the La Defense complex of offices, shops and apartments, which has become the largest single office development in Europe, employing 50 000 people and providing homes for over 20 000. The Palais Omnisports stadium at Bercy is the centre of a project including housing and a large new public park.

Level 1 (1–2 marks): *Simple description of new housing developments replacing slums.*

Level 2 (3–4 marks): *More detailed description, with supporting examples.*

Level 3 (5–6 marks): *Fully developed answer describing the main features of at least one major redevelopment project with an awareness that some older residential districts have been improved, rather than simply cleared.*

4 Describe the improvements made to Paris' transport infrastructure since 1970. *(6)*

Major improvements include the completion of an orbital motorway, the Boulevard Peripherique. This has allowed inter-regional traffic to by-pass Paris. Within Paris, more car parking spaces have been provided. The Metro system has been improved and modernised. A new, high speed Metro, the RER, has been built. This extends far beyond the original Metro system into the outlying suburbs of Paris, thus reducing the need for commuters to travel by road. The RER also extends to EuroDisney, successfully keeping much tourist traffic off the roads. The RER connects with the old Metro at some stations, allowing for integration between the two systems. A new, fully automated Metro line, called the Meteor, has been built under central Paris.

Improve Your Mark!

Level 1 (1–2 marks): *General references to road improvements and/or improvements to the Metro.*
Level 2 (3–4 marks): *More detail on the improvements.*
Level 3 (5–6 marks): *Demonstrates a clear understanding of the importance of the improvements to the Metro and the development of the RER systems.*

CHAPTER 5 deals with urban change in developing countries. This is covered under four sections: growth of world cities, urban growth, differences in residential areas including self-help schemes and problems in developing cities, using Kolkata (Calcutta) as an example. An additional case study investigates Rio de Janeiro as an example of urban problems and solutions. Other studies in the chapter include housing contrasts and improvements in São Paulo and urban problems in Kolkata.

Growth of World Cities   📖 78 – 79

The focus here is on changes since 1950 and especially the rapid growth of very large cities in developing countries. Extensive use is made of graphs, maps and tables of statistics.

Activity Sheet 5.1 provides further statistics and requires pupils to produce a bar graph to illustrate them (ideally, this should be done in a suitable spreadsheet package).

Urban Growth   📖 80 – 81

The familiar, but very important, list of push and pull factors explaining the movement of people from the countryside to the cities dominates this section. Figure 5.7 on page 80 is an excellent resource, summarising the main factors. The focus then shifts to urban land use in developing world cities. Figures 5.9 and 5.10 on page 81 should be compared and contrasted with Figure 3.19 on page 42 and Figure 3.21 on page 43.

Activity Sheet 5.2 provides a useful homework or revision exercise, requiring pupils to complete a matrix of push and pull factors. This provides a good link to the subject of migration in Chapter 2.

Activity Sheet 5.3 uses an unfinished copy of Figure 5.10 on page 81 as a starting point to test pupils' understanding of the differences in urban land use and functional zones in cities in developing and developed countries.

Differences in Residential Areas   📖 82 – 83

Using the example of São Paulo in Brazil, this section examines the extreme contrast in living conditions of the rich and the poor. Figure 5.11 on page 82, showing the wealthy housing in São Paulo, and the accompanying paragraph are useful contrasts to the perceived image of housing conditions in cities in developing countries. However, the main focus of the section is on the housing problems facing the poor. There is an excellent description of a shanty town on page 82. The value of self-help schemes is well covered on page 83, with accompanying Figures 5.13 and 5.14.

Activity Sheet 5.4 focuses on the development of a shanty town through time, as it transforms itself into a well-built residential area through a self-help scheme. This activity will stretch more able pupils.

Problems in Developing Cities   📖 84 – 85

The example of Kolkata is used to show the extent of the problems facing cities in developing countries. It is important to emphasise the point made in the opening paragraph on page 84: '…it is easy to provide a negative stereotype of life in a developing city. However … the authorities have made real improvements in the urban environment.' Direct pupils to the following website www.kolkatabeckons.com so that they gain a more positive view of life in Kolkata. The following gives a taste: 'Fascinating, bewitching, bewildering, that's Kolkata … a seething mass of activity with a cosmopolitan atmosphere. Kolkata has become a busy and flourishing city, the centre of cultural as well as political and economic life of Bengal. It is a city that throbs with vibrant life forces.'

Activity Sheet 5.5 concentrates on the more positive aspects of Kolkata by describing the measures taken to tackle its urban problems. There are some hopeful statistics to support the claims that progress is being made. It is worth pointing out to pupils that both the Kolkata Municipal Authority and the Metropolitan Development Authority are government agencies, which will try to paint as rosy a picture as possible. Nonetheless, this is a more upbeat message than that normally associated with Kolkata.

Case Study 5 Rio de Janeiro   📖 86 – 88

Rio de Janeiro provides a detailed case study of the problems facing major urban areas in developing countries and of some effective solutions, which include the self-help housing scheme in Roçinha and the city authority's Favela Bairro project. The new town of Barra da Tijuca provides an interesting study of a new town built to house Rio's wealthy inhabitants who have sought more space and a safer place to live. A self-contained town of luxurious apartments, Barra has its own shanty towns to house the low-paid cleaners, cooks and housekeepers employed by the wealthy – a suitably ironic note on which to end a study of contrasts.

Case Study Extra Agadir, Morocco

This case study focuses on the project to improve the shanty town of R'Mel in Agadir. The importance of working with the community's support and of flexible government plans is emphasised.

1 Study Figure A.

a) Draw a bar graph to illustrate the statistics in Figure A. Use a suitable spreadsheet package if possible, but work on paper if necessary. *(10)*

b) Which continent shows (i) the fastest and (ii) the slowest percentage growth between 1970 and 1995? Explain the differences in percentage growth. *(5)*

c) How does the actual and predicted increase in the percentage of population in urban areas vary between Africa and Europe? Explain the differences. *(4)*

Continent	Africa			Asia			Europe			North America			South America		
Year	1970	1995	2025	1970	1995	2025	1970	1995	2025	1970	1995	2025	1970	1995	2025
% of population in urban areas	22	35	52	23	32	53	62	71	81	70	75	82	54	72	80

A The percentage of the population in each major continent that live in urban areas.

2 Figure B shows the percentage of the world's 100 largest cities situated in each continent.

a) Complete the final row of the table, which shows the change between 1990 and 2002. *(2)*

b) Name the largest cities in each of the six continents, using Figure 5.3 on page 79 of the pupil book to help you. *(3)*

c) Which continent shows (i) the greatest increase and (ii) the greatest decrease in numbers? Explain why in both cases. *(6)*

Continent	Africa	Asia	Europe	North America	Oceania	South America
% in 1990	6	44	19	13	2	16
% in 2002	10	42	13	16	2	17
Change between 1990 and 2002	+ 4	- 2				

B Percentage of the world's 100 largest cities that are situated in each continent.

1 Complete the table below by giving reasons why people in developing countries are attracted to the cities and leave the countryside. Use Figure A to help you. *(24)*

Reasons	Employment opportunities	Pressure on the land	Over-population	Mechanisation	Services	Natural hazards
Push						
Pull						

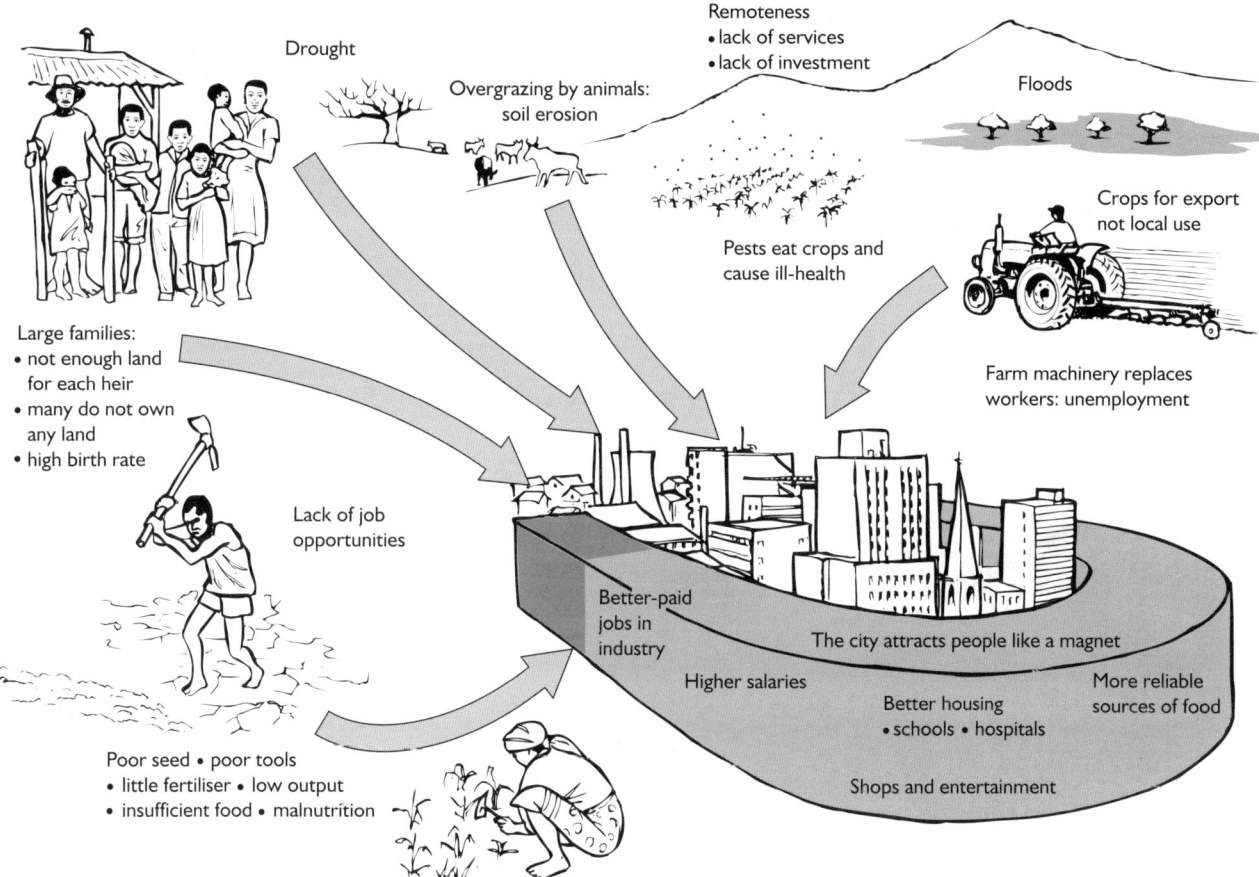

A Rural push factors and urban pull factors.

2 How does the reality of life in a developing city compare with what rural migrants expect to find? *(6)*

1 Study Figure A, which is based on Figure 5.10 in the pupil book.

  a) Complete the unfinished key for sectors A and B.   *(3)*

  b) What do you understand by the terms (i) *favela* and (ii) *periferia*?   *(4)*

  c) Why do many new migrants make temporary shelters when they arrive in the city, rather than rent or buy a home?   *(1)*

  d) How does this model differ from the model for cities in developed countries?   *(4)*

  e) Why are the modern factories in Brazilian cities built in narrow sectors?   *(1)*

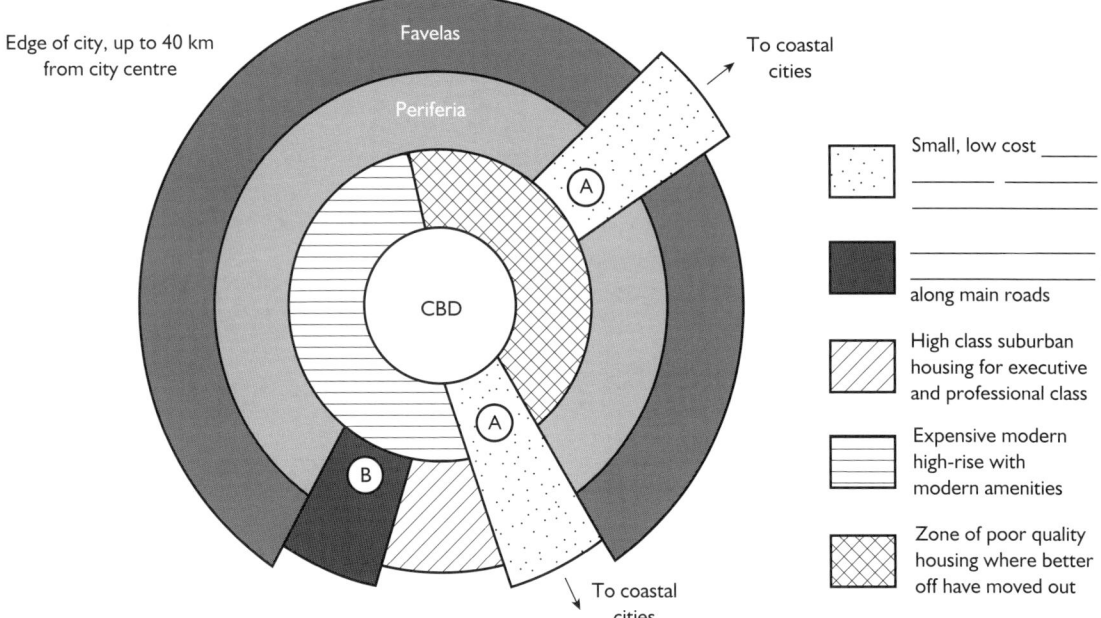

A Urban Land Use Model for a city in the developing world (based on cities in Brazil).

2 Name a city in the developing world you have studied. Compare and contrast the development of that city with the model shown in Figure A.   *(5)*

3 Describe how and account for the fact that the residential areas in developing world cities are very different for the rich and the poor.   *(5)*

In the past, shanty towns were thought to be places of squalor and despair, populated by desperate people who posed a threat to national security. This idea was accepted by many governments, who bulldozed or torched the shanty towns and displaced the inhabitants. However more recently, shanty towns have been recognised to be much more hopeful and positive places than previously thought. The reason for this is that the negative image of the shanty town is based on short-term observation. Once a longer term view is taken, it becomes clear that shanty towns pass through a series of stages of development. The early stages do indeed seem to be squalid and chaotic, but they are certainly not hopeless, and the later stages may see the development of well-serviced, well-built suburbs.

As time passes, the shanty town dwellers are able to buy better building materials and improve their homes. Thus, a shack can be transformed into a well-built, fully-serviced house in a process that may take 20 years. The squatters are capable of building homes for themselves, but they cannot provide the infrastructure. Some governments, the World Bank and leading charities have funded site and services schemes, where the authorities provide a site, basic building materials and links to electricity, water and drainage. Roads, clinics and schools may eventually be built as the site develops. In this way, the shanty towns are improved and a strong community spirit can be created.

1 a) Why did many governments in the past oppose the development of shanty towns? (3)

   b) Why have opinions about shanty towns changed? (4)

2 a) Explain what is meant by *site and services schemes*. (4)

   b) Why have site and services schemes become popular with governments? (2)

3 Use Figures A and B to help you describe and explain community self-help schemes in São Paulo. (5)

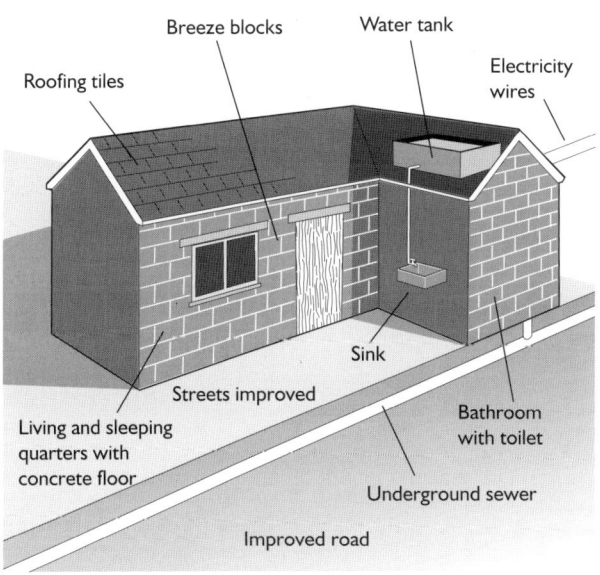

A A self-help housing scheme in São Paulo.

B A community self-help housing project in São Paulo – a 'slum of hope'.

The problems faced by the Indian city of Kolkata (Calcutta) are summarised on pages 84–85 of the pupil book. The book states that 'despite the poverty and squalor, and even with the scale and range of the problems facing the city, and its lack of resources, the authorities have made real improvements in the urban environment'. This activity sheet looks at some of those improvements.

For 30 years, the Kolkata Metropolitan Development Authority (KMDA) has been playing a key role in the planning and development of urban infrastructure within Kolkata Metropolitan Area. Infrastructural improvements have included widening and improvement of existing roads; construction of expressways, bypasses, subways, flyovers and bridges; construction of bus terminals; improvement of street lighting and traffic signals, and India's first underground railway system, the Metro Railway, which runs from DumDum to Tollygunge. There are 17 stations, of which 15 are underground. An extension from Tollygunge to Garia, which will add another seven stations, is under construction.

The Second Hooghly Bridge, with its network of criss-crossing and overlapping flyovers, is a marvel in its own right. This new toll bridge gives a fast connection from central Kolkata to its twin city of Howrah, reducing the traffic congestion suffered by the original Hooghly Bridge.

Kolkata's most highly publicised problems involve its shanty towns, or bustees. The website of the Kolkata Municipal Authority takes an upbeat view when it states:

'The unhealthy atmosphere of the slums is no more. Paved roads, street lights, piped drinking water, and an organised sewer system have made a great improvement in the slums.'

The KMDA has undertaken several bustee improvement schemes since 1985, which have resulted in considerable improvements, including:

- construction of roads and passages; more than 600 km of paved roads have been constructed

- the provision of more than 20 000 water tap points and connections, 15 000 street lighting points and 45 000 permanent sanitary latrines

- the drainage and sewer systems which have been revamped and renovated in these areas; 19 km of drains have been repaired and rehabilitated and almost 75 km of new drains and sewers have been installed

- the expansion and improvement of the water supply (see Figure A)

Year	1985	1990	1995	2000
Water (million litres)	818	1027	1091	1146

A Volume of daily treated (potable) water supplied from 1985–2000.

- almost 900 people working in health programmes, which have helped more than 700 000 slum dwellers; the infant mortality rate for Kolkata (43 per thousand) is much lower than the all-India average of 84 per thousand

- improvements in infant mortality rates which have been noted in the areas covered by two large projects conducted in Kolkata during the 1990s (see Figure B)

Project	Infant mortality rate in 1992	Infant mortality rate in 1999	Crude birth rate in 1992	Crude birth rate in 1999
Kolkata Slum Improvement Programme	43	16	23	9
India Population Project – Eight	56	25	20	13

B The effects of Kolkata projects on infant mortality rates in the 1990s.

- the passing of a Tenancy Act whereby the rights of the slum dwellers were recognised and ensured; they can no longer be evicted by their landlords.

1 Using the information provided on this activity sheet, describe and explain how the urban problems facing Kolkata are being tackled. What evidence is there that the policies are achieving some success? *(8)*

2 Using the Internet, visit the website www.kolkatabeckons.com to help you write an article on the attractions of Kolkata. *(10)*

Agadir, Morocco

Agadir is one of the largest cities in Morocco with a population of nearly half a million. Figure A shows its location. It has a serious problem of substandard housing and particularly shanty towns (or bidonvilles); by 1992, there were 77 separate shanty areas with some 12 500 homes. The city was badly damaged by an earthquake in 1960 and many of the residents of the bidonvilles have lived in their shanties for decades, often since the earthquake.

A The location of Agadir and R'Mel, Morocco.

R'Mel is a bidonville in the Inezgane commune of Agadir (see Figure A). An established squatter community, R'Mel is the scene of an integrated improvement project by the Moroccan National Shelter Upgrading Agency (the ANHI). The aim of the agency is to improve the living conditions of over one thousand lower income households in R'Mel. The project is based on the following:

- help for people to purchase sites on which to build permanent housing

- provision of sites with the basic services of water, sewer, electricity and streets

- self-help principles with active community participation, which result in new homes being built and existing ones improved.

The Development of the Project

The project was initially designed by the ANHI in 1988, providing for the gradual demolition of the shanties and for moving people as new sites were completed and first payments were received. However, the project immediately ran into difficulties because about 40% of the lower income households did not live in shanties. They had replaced their shanties with small houses over the years and they refused to have their homes demolished. The project had to be redesigned to allow for building an infrastructure around the existing homes. This slowed the scheme down and increased its cost. The local authority was not pleased because they wanted the appearance of the area to be greatly improved. However, the authority eventually agreed to the redesigned project. To help cover the increased costs, additional lots for larger houses and small apartment blocks were added at the edge of the site, for sale at a profit.

The redesigned project had much greater community involvement. Group meetings, publicity campaigns and individual advice sessions were organised. Families were given assistance with demolition and removal, and provided with technical advice by the on-site architect, engineer and other staff. The community worked together on the project. Those who built their new homes first learned construction skills, which they then used to work on their neighbours' houses. It took 14 months to demolish all the shanties. Of course, not every family could afford the price of the land, even if it was subsidised. The poorest were allowed to rebuild their shacks, using improved materials and with utility connections.

The Results

■ Four thousand new homes were built.

■ 650 shanties were improved and connected to utilities.

■ 400 additional low income houses were connected to utilities.

■ Jobs, estimated at 25 000 per year, were created during the three-year construction phase.

■ The local authority built schools, shops and markets within R'Mel and a mosque, a health centre and a police station nearby. The former shanty town became indistinguishable from other more established areas of Agadir.

■ The project improved R'Mel in a sustainable manner. Nobody was forced to leave, so the community was preserved. Residents improved their living standards through their own efforts and gained a voice in local decision-making.

Exam Practice Questions

1 What is a shanty town? *(2)*

2 Why do shanty towns develop? *(3)*

3 How can shanty towns be improved? Use a named example or examples. *(7)*

Agadir, Morocco

This model answer is designed to show you how to get full marks for the longer question in the Case Study Extra for this section. Read the question carefully and write your own answer. Then read the model answer and the examiner's notes to see what he or she is looking for when awarding the marks. Decide how many marks you think the examiner would give your answer. Decide how to change your answer to this particular question to increase your marks. What will you do differently next time you answer a similar question?

3 How can shanty towns be improved? Use a named example or examples. *(7)*

> Improvement of shanty towns involves providing basic amenities and services, and improving or replacing the shanties. R'Mel is a shanty town in Agadir, Morocco. A project to upgrade the area assisted people to buy sites on which to build new permanent homes – this sorted out the land ownership issue. The sites were provided with amenities. Families were given assistance with demolition and removal and provided with technical advice by the on-site architect, engineer and other staff; however they completed the work themselves with help from other members of the community. The local authority built schools, shops and a market in the area. Four thousand new homes were built and over a thousand existing homes improved and provided with amenities. The former shanty town became indistinguishable from other suburbs of Agadir. The R'Mel project is an example of a self-help project.

Level 1 (1–2 marks): *Basic answer including reference to improvements to buildings and provision of basic amenities.*

Level 2 (3–4 marks): *Greater detail on the improvements, using a named example.*

Level 3 (5–7 marks): *Detailed answer including a named example, demonstrating clear understanding of the partnership between government and individuals, and the nature of self-help.*

6 EMPLOYMENT STRUCTURES

CHAPTER 6 is a short one, dealing with employment structures and development. This is covered under five sections: classification of economic activities, employment structures and changes over time, differences between places and employment structures and development. The chapter also includes an explanation of triangular and percentage bar graphs.

Classification of Economic Activities   📖 92

The four-fold classification from primary to quaternary is used. A useful extra exercise here would be to ask pupils to list the economic activities that went into producing the pupil book and to classify those activities into the correct groups. Primary activities include the foresters who grew the trees to make the paper on which the book is printed; secondary activities include the production of the paper; tertiary activities include the work of the editorial and office staff employed by the publisher and the lorry drivers delivering the books to the school. Quaternary activities might include the IT specialists producing the computers and software on which the book's layout was designed.

Employment Structures and Changes over Time   📖 92

The inexorable move from an employment structure dominated by the primary sector to one dominated by the tertiary sector is clearly shown in Figure 6.2 of the pupil book.

Activity Sheet 6.1 includes a table of statistics showing how the structure of the UK's employment changed rapidly between 1982 and 2002. Here, the missing words in Question 4, a fill-in-the blanks exercise, are in italics. 'The decline in employment in mining, allied with a decline in *farming* and fishing, led to a halving of employment in *primary* industry. Secondary industry employment also declined as labour-intensive *manufacturing* firms such as steel and *textiles* closed or reduced their labour force as a result of increasing *automation*. The growth of tertiary employment is remarkable, especially in the *financial* sector.'

Differences between Places   📖 93

Large and small-scale differences between places are shown by the differences in the UK's regions in Figure 6.3 and in different types of town in Figure 6.4 on page 93. This section also shows pupils how to interpret a triangular graph.

Activity Sheet 6.2 supports page 93 and also tests pupils' skills with graphs. It uses Figure 6.3, requiring pupils to use protractors to read the percentage of the total workforce employed in each employment sector in five UK regions. The statistics are used to complete a table and then the statistics for three regions are placed on a triangular graph.

Employment Structures and Development   📖 94

Figure 6.5 on page 94 clearly shows the contrasting employment structures of sixteen of the richest and poorest countries in the form of a map. Figure 6.6 shows the same sixteen countries' statistics presented in the form of a percentage bar graph. It is useful to ask pupils which works better in Question 3 on Activity Sheet 6.3.

Activity Sheet 6.3 requires pupils to identify the employment structures of four countries from unnamed pie graphs. Graph A is the UK, B is Nigeria, C is Brazil and D is South Korea.

1 a) Tick the correct industry sector for each of the following statements. (4)

	Primary	Secondary	Tertiary	Quaternary
High-tech industries providing information and expertise				
Industries providing a service				
Industries involved in the manufacture of goods				
Industries taking raw materials directly from the earth				

b) Explain how primary and secondary industries are linked. (2)

2 Place a letter in each box below to describe the correct industry sector (P = Primary, S = Secondary, T = Tertiary, Q = Quaternary) for each of the following industries:

- education ☐
- steelmaking ☐
- car assembly ☐
- textiles ☐

- health ☐
- retailing ☐
- mining ☐
- forestry ☐

- office work ☐
- farming ☐
- micro-electronics ☐
- papermaking ☐

- software design ☐

3 Study Figure A, which shows the number of people employed in the UK in each sector of industry.

a) Draw pie graphs showing the total percentages employed in the three sectors in 1982, 1992 and 2002. (6)

b) How did the percentages employed in primary and secondary industry change between 1982 and 2002? Why was this? (6)

c) Which sub-sector of employment showed the greatest increase between 1982 and 2002? Why was this? (3)

Sector	1982	%	1992	%	2002	%
Agriculture, hunting, forestry & fishing	679	2.6	624	2.3	432	1.5
Mining & quarrying	560	2.2	337	1.2	188	0.6
Total primary	**1239**	**4.8**	**961**	**3.5**	**620**	**2.1**
Manufacturing	5375	20.8	4523	16.6	3941	13.4
Construction	1872	7.2	1922	7.1	1953	6.6
Total secondary	**7247**	**28.0**	**6445**	**23.7**	**5894**	**20.0**
Distribution, hotels & restaurants	5696	22.0	6218	22.8	6795	23.0
Transport, storage & communication	1596	6.2	1617	5.9	1765	6.0
Banking, finance & insurance	3273	12.6	4306	15.8	5675	19.2
Public administration, education & health	5625	21.7	6358	23.4	6995	23.7
Other services	1218	4.7	1344	4.9	1774	6.0
Total tertiary	**17 408**	**67.2**	**19 843**	**72.8**	**23 004**	**77.9**
Grand total	**25 894**	**100**	**27 249**	**100**	**29 518**	**100**

A Employment structure in the UK in 1982, 1992 and 2002.

4 The following is a commentary on Figure A. Fill in the blanks with suitable words chosen from the word bank beneath the passage.

The decline in employment in mining, allied with a decline in _____ and fishing, led to a halving of employment in _____ industry. Secondary industry employment also declined as labour-intensive _____ firms such as steel and _____ closed or reduced their labour force as a result of increasing _____ . The growth of tertiary employment is remarkable, especially in the _____ sector.

- manufacturing
- farming
- textiles
- primary
- financial
- automation

Figure A shows the employment structures for the regions of the UK.

UK average

Primary 2%
Secondary 26%
Tertiary 72%

Figures give percentage
of total employed population
in each region

8 million
5 million
1 million
0.5 million

Area of circle proportional to
total employed population
in each region

Region	Primary	Secondary	Tertiary
East Anglia	8	26	66
East Midlands	9	33	58
North	6	29	65
Northern Ireland	10	19	71
North-west	1	32	67
Scotland	4	26	70
South-east			
South-west			
Wales			
West Midlands			
Yorkshire & Humberside	7	32	61
UK Average	**2**	**26**	**72**

0   100 km

N

Scotland

Northern Ireland

North

Yorkshire and
Humberside

North West

East Midlands

Wales

East Anglia

West Midlands

South East

South West

A Employment structure for the regions of the UK.

1 a) Using Figure A, complete the table above
   right. (6)

 b) Which region has the highest percentage of its
   labour force in the secondary sector?
   Why is this? (3)

 c) Which region has the highest percentage of its
   labour force in the tertiary sector?
   Why is this? (3)

2 Which two regions have the largest employed
  populations? Why is this? (3)

3 Plot the statistics for the south-east, the West
  Midlands and Northern Ireland on the triangular
  graph, Figure B. (3)

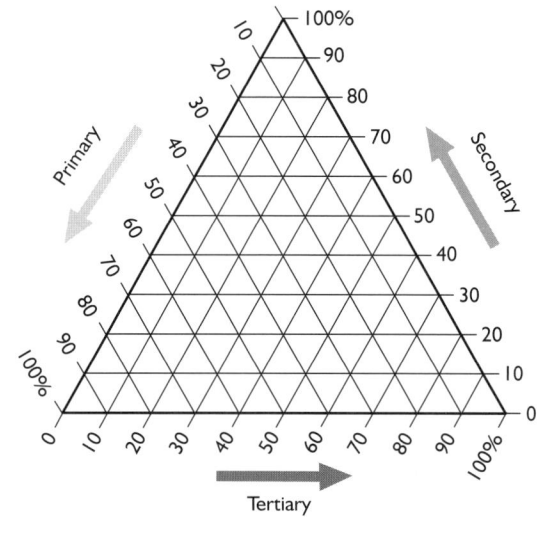

B Triangular graph.

The New Wider World (Second Edition): Teacher's Resource Book
Neil Punnett and Alison Rae © Nelson Thornes 2003

Figure A shows the employment structures in four countries – Nigeria, Brazil, the UK and South Korea.

1 Label each pie graph with the name of the correct country, giving reasons for your choices. *(8)*

2 In general terms, what features of a country's employment structure indicate that it is
(i) a developed or (ii) a developing country? *(4)*

3 Study Figures 6.5 and 6.6 on page 94 of the pupil book.

a) What relationship have the two figures been drawn to show? *(2)*

b) Which figure is, in your opinion, more effective at showing the relationship?
Give reasons for your answer. *(4)*

(i)
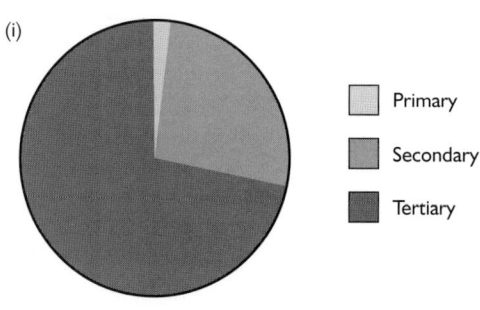

- Primary
- Secondary
- Tertiary

(ii)
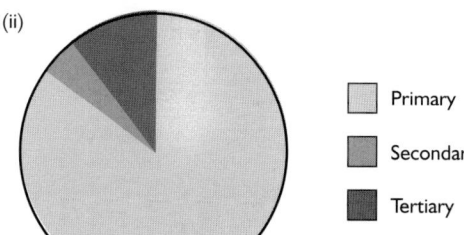

- Primary
- Secondary
- Tertiary

(iii)
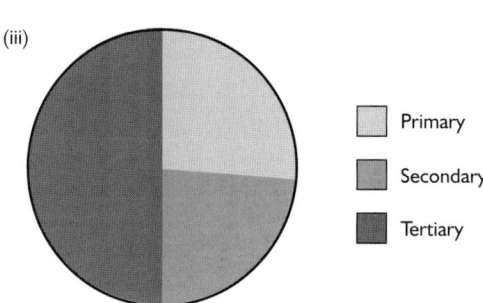

- Primary
- Secondary
- Tertiary

(iv)
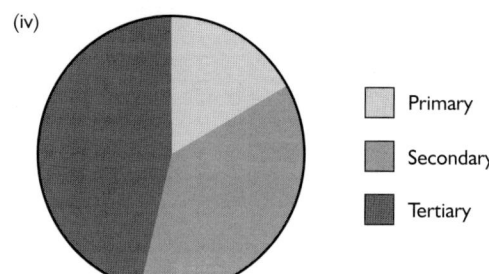

- Primary
- Secondary
- Tertiary

A Employment structure in four countries.

CHAPTER 7 strikes a balance between the theoretical basis of farming, i.e. the systems approach, and the way the theory is applied in the real world in both MEDCs and LEDCs. A particular emphasis is placed on the UK to match the specification. The chapter is divided into seven sections on the main GCSE and Standard Grade topics, with a case study on India.

Farming Systems and Types   📖 96 – 97

The systems approach is crucial to farming studies at this level and is illustrated on page 96. Classification of farming is also considered.

Activity Sheet 7.1 requires pupils to consider all parts of the systems flow diagram.

Activity Sheets 7.2 and 7.3 both explore the classification of farming.

Factors affecting Farming in the UK   📖 98

These are divided, as is common, into physical, human and economic, and political factors, providing links to the section on the Common Agricultural Policy (CAP) on page 107 and aspects of the other EU examples, such as East Anglia on page 101 and Denmark on pages 102–103. The different factors affecting a range of farming types could be the basis for classroom discussion.

Distribution of the Main Types of Farming in the UK   📖 99

The basic classification shown in Figure 7.6 on page 99 is a good place to start. The reasons for the distribution, which are mainly physical, are explored. The photographs will help pupils visualise landscapes which may not be familiar to them.

MEDC Farming Case Studies   📖 100 – 105

In any GCSE or Standard Grade course, a broad range of case studies is required. This section provides studies on various farming types and locations – pastoral farming in the Lake District, arable farming in East Anglia, mixed farming in Denmark and a comparison of Mediterranean farming in the Mezzogiorno before and after the massive input of government and EU funding. Activity Sheets 7.3 and 7.4 provide work based on EU farm types.

Activity Sheet 7.3 shows a photograph of a landscape that includes set aside land, which ties in with page 107 on CAP. This is an activity sheet that develops field sketching and annotation skills, and so may support field studies.

Activity Sheet 7.4 considers dairy farming in the UK and links with both physical factors and the impact of CAP. A question concerning the distribution of dairy farming allows pupils to practise their descriptive skills. Construction and interpretation of climate graphs are also included here.

Changes in Farming   📖 106 – 107

Changes caused by intensification, CAP and economic pressures are considered. Pupils from rural areas may have first hand experience of the impact of these changes. Effects on the landscape could usefully be discussed. We all shop in supermarkets, so their control over the farming industry today could also be turned into a controversial debate.

Farming and the Environment   📖 108 – 109

The use of chemicals, habitat issues, organic farming and genetically-modified (GM) crops are included. Pupils will probably be interested to research these topics themselves. Useful websites are: www.soilassociation.org, www.foe.co.uk and www.nfu.org.

Food Supply and Malnutrition   📖 110 – 111

In contrast to the MEDC situation, this section considers a more negative side of agriculture, i.e. when yields prove insufficient to feed a population. Dietary energy supply (DES) is a recent way of quantifying this problem.

Case Study 7 Subsistence Rice Farming in the Lower Ganges Valley   📖 112 – 113

This case study describes rice farming as a type of subsistence agriculture, emphasising recent changes.

Activity Sheet 7.5 focuses on a contrasting commercial LEDC cash crop system, plantation coffee production. In particular, it considers the relationship between the producer and the consumer countries.

Case Study Extra Market Gardening with Fruit Farming and Flower Production

This topic adds another type of farming and location to the range of case studies offered, using the Fylde area of Lancashire, a less well-known farming district, but nevertheless an excellent example of market gardening in the UK.

The farm system has inputs, processes and outputs.

1 Complete the boxes on Figure A, using the words in the panel.
One has been done to give you a start.   (4)

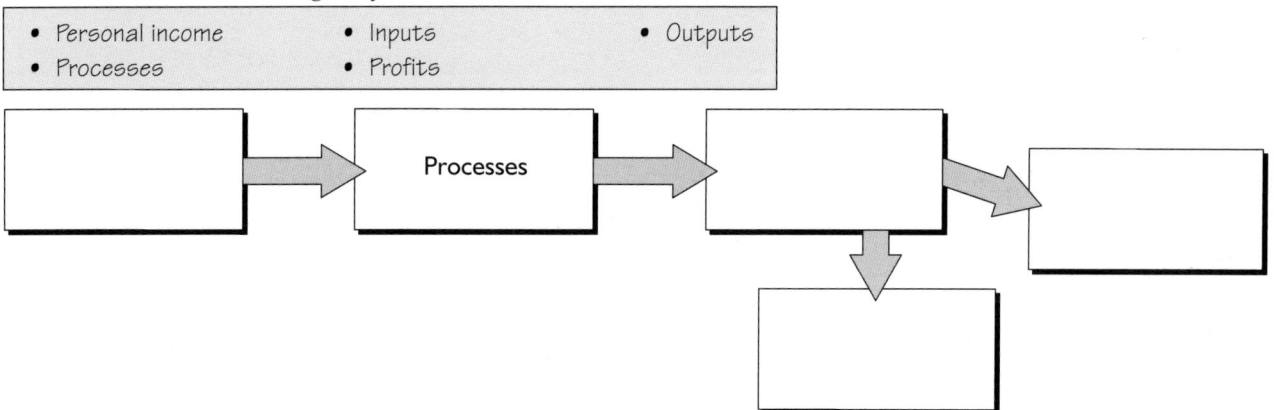

- Personal income   • Inputs   • Outputs
- Processes   • Profits

Processes

A The farm system.

2 Inputs into the farming system fall into two categories, physical or natural inputs and human, economic, social or political inputs.

a) Write a definition for each of these two types of input.   (2)

b) (i) In Figure B, shade the physical inputs box in green and the human inputs box in red.

(ii) Shade each of the words in the centre in either red or green, according to their category, and draw a line to link each to the correct box. Two have been done for you.   (9)

3 Physical inputs in a farming system interlink with each other, i.e. each affects one or more of the others.

a) Add arrows to Figure C to show these links. Some have already been done for you.   (5)

b) Choose two of the links and explain how one factor affects the other.   (4)

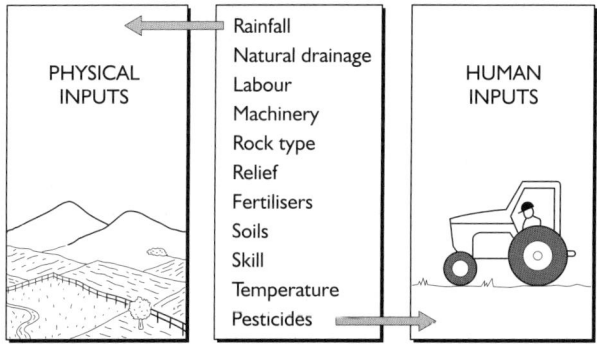

B Physical and human inputs in the farming system.

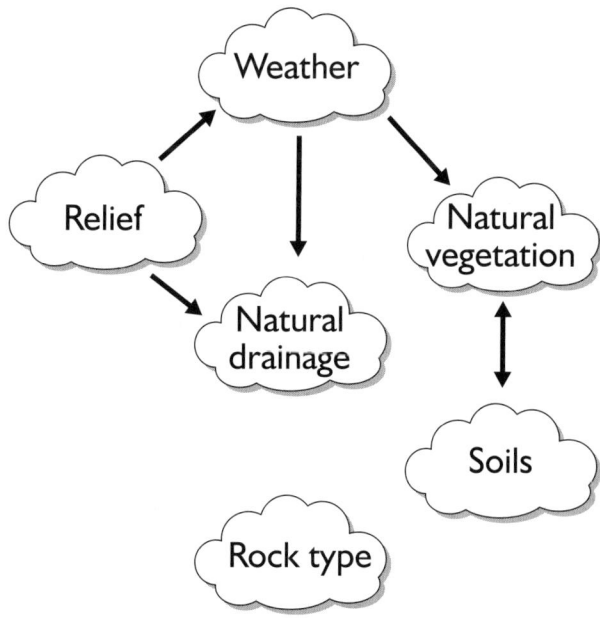

C Physical inputs in the farming system.

Every farm has to fit into one of the categories in each set below.

Set A: Arable **or** Pastoral **or** Mixed.

Set B: Commercial **or** Subsistence.

Set C: Intensive **or** Extensive.

Set D: Sedentary farming **or** Shifting cultivation.

Set E: Privately owned **or** Rented.

1 Write a brief definition for each of the terms (there are eleven of them).　　　　(11)

2 Decide which category in each set each of the following farming types belongs to. All of the
types of farming are explained in more detail in the pupil book on the pages indicated.　　(6 x 5)

(i)　Sheep farming in the Lake District (page 100).

(ii)　Arable farming in East Anglia (page 101).

(iii)　Mixed farming in Denmark (pages 102–103).

(iv)　Farming in the Mezzogiorno in the 1960s (page 104).

(v)　Farming in the Mezzogiorno today (page 105).

(vi)　Rice farming in the Lower Ganges Valley (pages 112–113).

3 Commercial farming across the world has become increasingly intensive in recent years.

a) Explain how farmers make their production more intensive. Use the extract below
and page 108 in the pupil book to help you.　　　　(4)

b) What do the extract and Figure A suggest about the intensification of agriculture?　　(3)

c) Discuss the advantages and disadvantages of the intensification of agriculture.　　(9)

Grazing animals convert grass … into edible meat and milk. Pigs and poultry were traditionally fed on the waste from human food. However, modern intensive farming uses grain, oilseed and fishmeal products to feed livestock. It takes 4–7 kg of grain to produce 1 kg of pork, and at least 15 kg of grain to get 1kg of beefsteak or hamburger.

The pharmaceutical industry is heavily involved. In the USA, livestock accounts for some 40% of total use of antibiotics – most of which are administered in feed.

Intensively-farmed diary cows produce ten times as much milk as their 19th century ancestors. Injections of bovine growth hormones (such as BST) can push yields even higher. But it is fair on the cow?

From **Understanding Global Issues**.

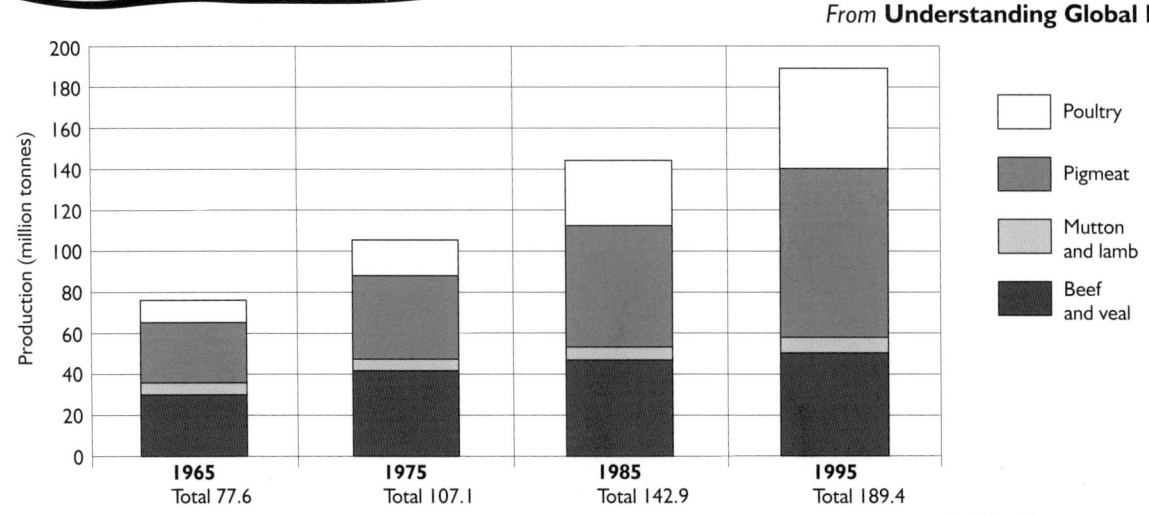

A World meat production.

Study Figures A and B.

1 a) Use the features on the photograph and the field sketch below to categorise the farm in as many ways as you can, in the same way as set out in 'Classification of types of farming' on page 97 of the pupil book. *(4)*

b) Suggest some of the inputs, outputs and processes that might be part of the system on this farm. Write your ideas in a systems diagram. *(6)*

A A farm in the Quercy region of south-west France.

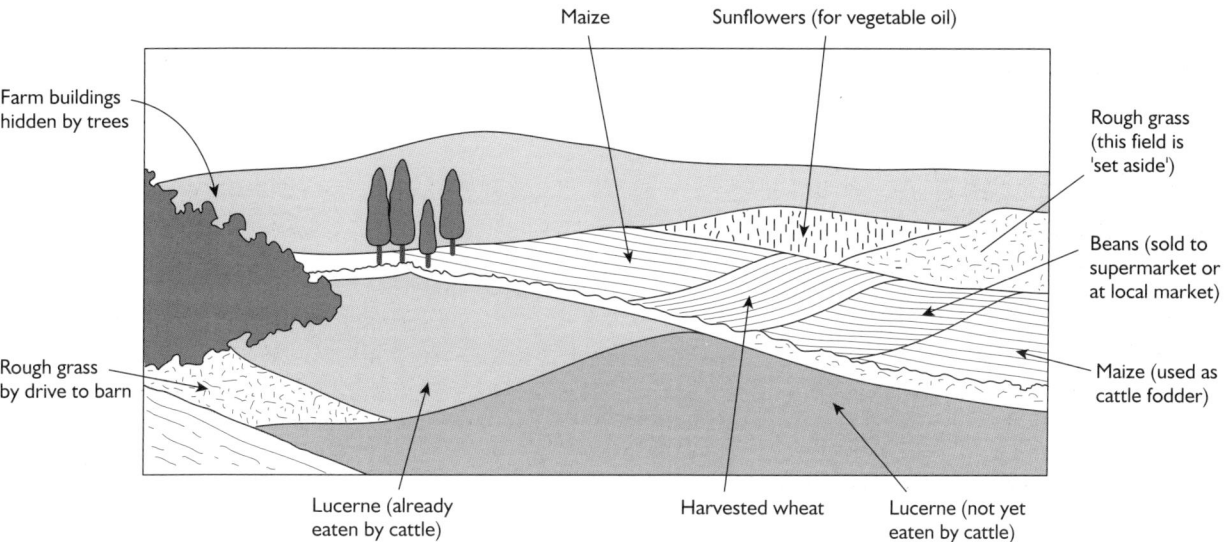

B Field sketch of a farm in the Quercy region of south-west France.

The aim of a field sketch is to show the important geographical features of a landscape. It can be based either on a photograph, as here, or on your own observation of the landscape. You do not need to be a good artist; it is the detail in the labels that is most important.

2 Draw two field sketches from the following sources.

a) A photograph of a farming landscape of your choice from Chapter 7 in the pupil book. *(5)*

b) Your own observation of a landscape near your home or one which you have visited on a field trip. *(5)*

Dairy farming is one type of cattle farming; the other is beef production. Some farms include both, because female livestock can be used for milk production, while males can be sold off for beef. However, usually cattle farms tend to specialise in one or other of these activities.

1 Study Figure A.

a) Name the seven dairy farming areas labelled 1 to 7 on the map. Choose from the regions named below: *(7)*

- around London
- Cornwall
- Somerset
- Cheshire Plain
- South Wales
- Devon
- Ayrshire and the Solway Firth.

b) Describe the distribution of dairy farming areas with reference to relief, climate and population. Your atlas will help you here. *(5)*

2 a) Complete Figure B, a climate graph for Somerset, using the statistics in Figure C. Remember that rainfall is shown using bars shaded in blue and temperature is given by a series of points joined together by a red line. The first two months have been done for you. *(5)*

b) Describe the main features of the Somerset climate. *(4)*

c) What features of this climate make the area particularly suitable for dairy farming? *(3)*

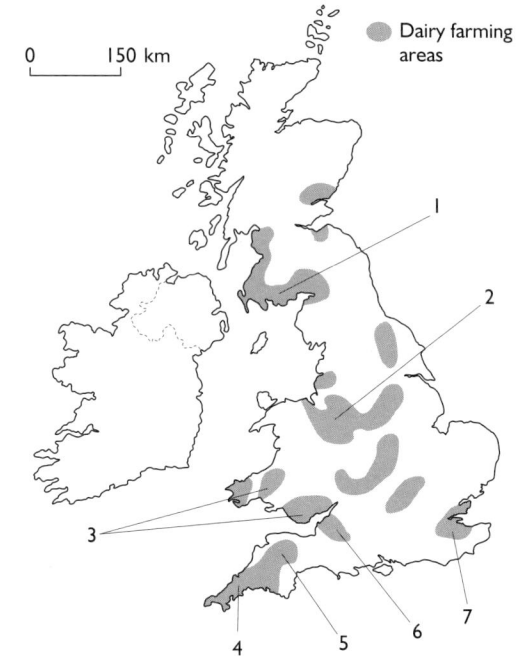

0   150 km

Dairy farming areas

A Dairying areas in the UK.

Month	J	F	M	A	M	J	J	A	S	O	N	D	
Temperature (°C)	4.6	4.7	6.6	9.2	11.2	14.9	16.9	16.4	14.5	11.5	7.7	5.7	
Rainfall (mm)	70	49	45	49	63	44	60	73	63	67	82	72	Total 737

C Average climate figures for the Somerset Plains, 23 m above sea level.

3 In what ways does the EU CAP influence dairying in the UK today? Hint: look at pages 106–107 in the pupil book. *(4)*

B Climate of the Somerset Plains.

Commercial plantations were first planted in LEDCs when they were colonies of European countries. Britain, for example, wanted to produce large quantities of some tropical crops. Plantations still exist today because they are one of the most efficient agricultural production systems in the tropics. The main feature of plantations is that only one crop is grown, but huge investment and skill is put into that one crop.

1 Plantation crops are largely exported by LEDCs as raw materials for the processing industries of MEDCs.

  a) Twelve countries are listed below. Put them into two separate lists of 'leading coffee producers' and 'leading coffee consumers'. *(12)*

• Brazil	• Mexico	• Japan	• Colombia
• Vietnam	• Kenya	• Italy	• Germany
• Indonesia	• UK	• France	• USA

  b) What do the countries in the list of producers have in common? *(2)*
  c) What do the countries in the list of consumers have in common? *(2)*

Production on plantations is an example of monoculture.

2 a) Explain what is meant by the term *monoculture*. *(2)*

  b) The newspaper extract below describes one particular problem of monoculture. Explain what is happening here. *(3)*

  c) List any more problems caused by monoculture. *(3)*

Like other Colombian growers, Mr Zuluaga has also been battling a plague of insects known as *broca*, barely visible to the human eye but deadly to the coffee bean. In the barn where he separates his beans according to quality, he has posted a warning placard which reads: 'Don't leave ripe or over-ripe fruit near your coffee fields. It feeds the *broca*. Attack the *broca* before it destroys your harvest.'

From **The Independent** 5 February 1997.

3 Study Figure A.

  a) (i) Who benefits most from coffee production? *(1)*

   (ii) Who benefits least from coffee production? *(1)*

  b) (i) What percentage of money would go to the MEDC? *(1)*

   (ii) What percentage of money would go to the LEDC? *(1)*

  c) Which people would have to share the 37.3% of the money generated by coffee production? *(3)*

  d) Comment on the fairness of the coffee production system. *(5)*

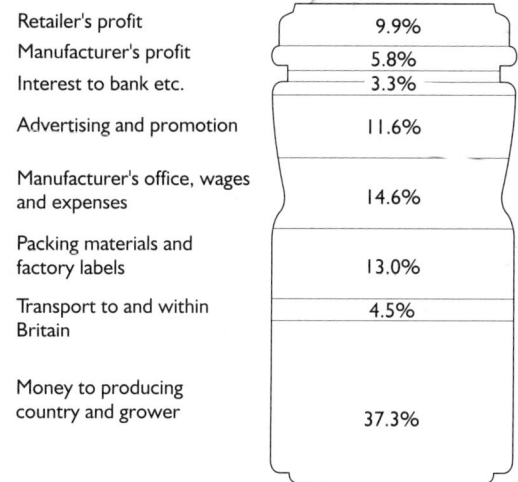

Retailer's profit	9.9%
Manufacturer's profit	5.8%
Interest to bank etc.	3.3%
Advertising and promotion	11.6%
Manufacturer's office, wages and expenses	14.6%
Packing materials and factory labels	13.0%
Transport to and within Britain	4.5%
Money to producing country and grower	37.3%

A Who gets what from the 'coffee jar'? Money generated by the coffee industry.

Market Gardening with Fruit Farming and Flower Production

Market gardening is probably the most intensive type of farming in the world. It has the highest inputs of labour and capital per hectare. Its main characteristics include:

- very high outputs – two or three crops of vegetables, fruit or flowers are harvested every year, even in Britain where the weather is often far from perfect

- increasing use of glasshouses, which protect crops from poor weather conditions, as well as allowing the farmer to create the best growing environment

- a lot of labour, some of which is skilled.

What is Market Gardening?

Market gardening is the intensive production of salad and other specialist vegetables, along with fruit and sometimes flowers. Typical crops are tomatoes, peppers, aubergines, courgettes, baby carrots, various lettuce, asparagus, strawberries, currants and loganberries. For example, Jersey and Guernsey, the two main Channel Islands, specialise in growing tomatoes and in flower production (see Figure A).

A Publicity material from 'Flying Flowers', Jersey.

Locations of Market Gardening

Figure B shows the key market gardening areas in the UK. There are two key points about the location of these market gardens.

- They must never be far from their customers as their crops are particularly perishable. They are easily bruised and damaged in transit, and have a very limited shelf life.

- Quality is everything and therefore produce should not travel far. The shortest possible time between production and consumption ensures maximum taste, greatest nutritional value and least waste. These are all things for which customers are willing to pay high prices. Customers are concentrated in urban areas, so market gardening often takes place on the rural-urban fringe. The area surrounding Greater London, which provides a concentrated market of over 10 million people, is therefore particularly important for this type of farming.

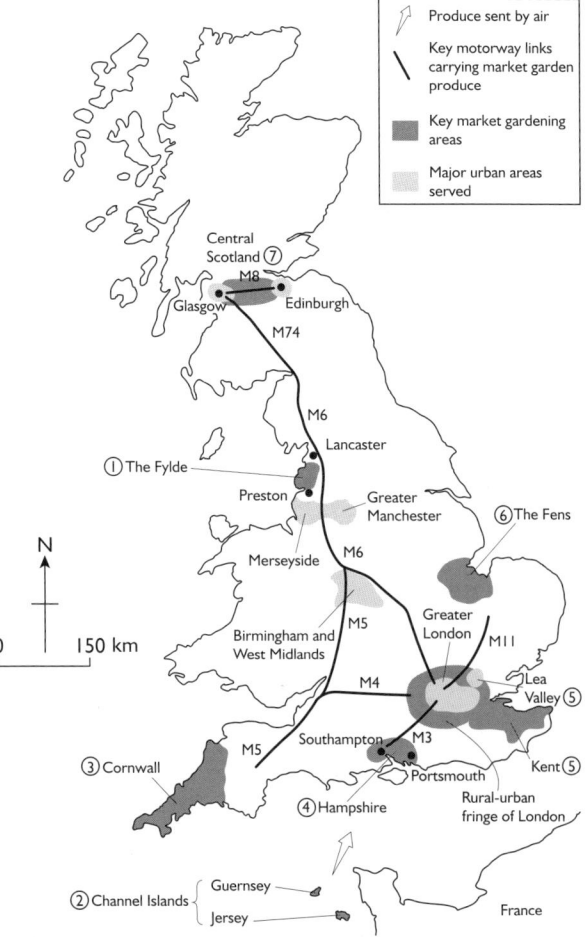

B Key market gardening areas in the UK.

Foreign Imports

However, Britain does import some of its market gardening produce. Spain is an important source of tomatoes and peppers. The major advantage of importing is that these normally seasonal products then reach our shelves all year round. Even before these imports from our EU partner were so important, the Channel Islands, part of the UK, sent its market garden produce to the mainland by air and still does so today. The problem of distance is overcome by careful packaging. Large quantities of materials are used to protect the goods, which add to the cost of them and are environmentally unfriendly.

Market Gardening in the Fylde Area of Lancashire

The Fylde is located in the north of the county of Lancashire between Lancaster and Preston (see Figures B and C). It has been a market gardening area for a long time, but its methods have changed and become increasingly high-tech.

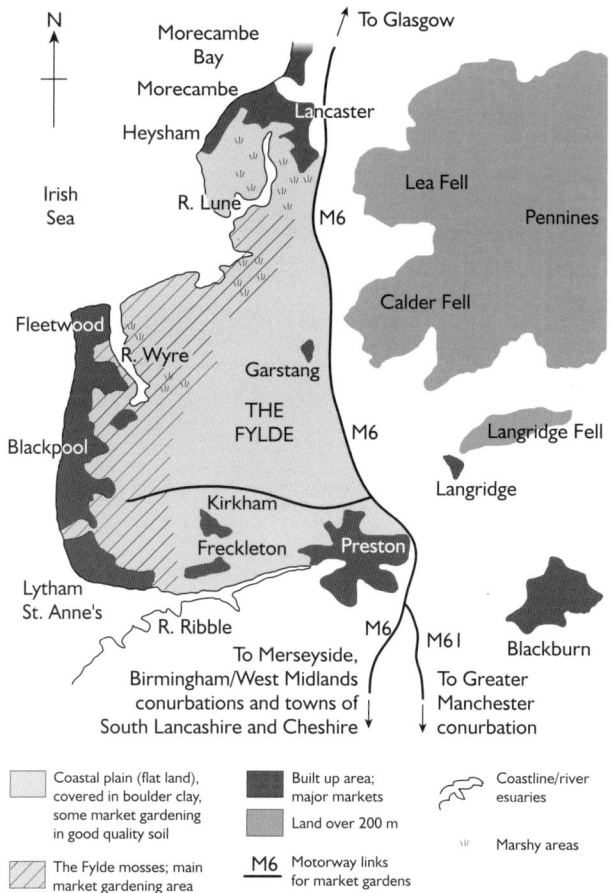

C The physical features and marketing links of the Fylde.

The original reasons for the development of market gardening in the Fylde were physical characteristics.

Temperature

The Fylde is a mild part of the country, with January temperatures of 4° C and July of 15° C. Extremes of temperature are rare and so the growing season is relatively long for this far north.

Precipitation

Rainfall levels are high, as the Fylde is in the west of the UK. The norm is 1000 mm per annum and this falls fairly evenly throughout the year.

Wind

One drawback is that winds can be strong, since there is little protection in the flat Fylde landscape from the prevailing westerly winds that come in from over the Atlantic. On the other hand, market garden crops, even if grown outside, are relatively low in height and so are less likely to be damaged by the wind.

Relief

The relief of the Fylde is another important physical factor in a farmer's decision about whether his land is suitable for market gardening. The areas shown in Figure B have one thing in common – they are all fairly flat. This is not important because there is a need for the use of large machinery as in most arable farming, but flatter areas often have better developed soils. Also the construction of glasshouses and poly-tunnels, which require a flat base, is easier. Most of the Fylde lies between zero and 50 m above sea level, making it particularly suitable. Higher parts are often drumlins, made of boulder clay left by glaciation, which can be easily avoided.

However, being flat and low lying, the Fylde mosses are prone to waterlogging. To prevent this, a system of drainage ditches has been constructed and these can easily be seen on any OS map of the area.

Soils

Soils in the Fylde are distinctive. They are called 'mosses' and are made of either boulder clay deposited by retreating glaciers during the last ice age or lowland peat. Both soils have high fertility and, where the two overlap, the minerals from both the clay and the organic matter in the peat make the soil particularly high in quality. The organic material lightens the heavy consistency of the clay, so that the soil does not become waterlogged, but remains sufficiently water retentive for the crops never to dry out. These soils are easy to work, which is a real advantage.

Modern Methods

An increasing amount of land is now used for cultivation under glass, so climate and soil conditions become less relevant. Salad crops are grown in this way. The soil can be altered with the addition of composts and fertilisers. Some growers even use hydroponics, growing crops just in water containing dissolved nutrients. Temperature and humidity can also both be controlled in a greenhouse environment. The costs of creating these manmade environments obviously increase the price the customer must pay for the produce, but losses due to weather, pests and diseases are reduced and that is a benefit.

Markets for the Fylde Produce

The Fylde is close to several urban areas. Lancaster is immediately to the north, Preston to the south, and beyond the towns of south Lancashire and the conurbations of Greater Manchester and Merseyside all represent nearby markets. Moreover, the region is linked by the M6 motorway to Birmingham and the West Midlands conurbation, and by the M6 and M74 to Glasgow. Both journeys are about two and a half hours long.

Exam Practice Questions

1 Describe the distribution of market gardening as seen in Figure B. *(5)*

2 Choose one of the market gardening areas shown in Figure B other than the Fylde or another area you have studied. Draw an annotated sketch map, similar to the one in Figure C, to show the physical and human factors affecting it. *(10)*

3 Imagine that you wish to set up a market gardening firm. Suggest four factors that you would consider about a particular location before setting up there. *(4)*

Market Gardening with Fruit Farming and Flower Production

These model answers are designed to show you how to get full marks for the longer questions in the Case Study Extra for this section. Read the question carefully and write your own answer. Then read the model answer and the examiner's notes to see what he or she is looking for when awarding the marks. Decide how many marks you think the examiner would give your answer. Decide how to change your answer to this particular question to increase your marks. What will you do differently next time you answer a similar question?

1 Describe the distribution of market gardening as seen in Figure B. (5)

Figure B shows the major market gardening areas in the UK. These areas are well scattered over the UK. They tend to be relatively near the coasts, as lower, flatter land is required. All have access to large urban areas via the national motorway network, which is an important factor in accessing markets. Most are also close to urban areas which act as markets; for example the region around London and the Fylde near Merseyside and Greater Manchester. More areas are located in the south rather than the north because the climate is more favourable.

> **Improve Your Mark!**
>
> Use the map really carefully and refer to it closely. Name market gardening areas and give examples – you get marks for naming specific areas. The other features, such as the motorways, are shown for a reason, so spot these and use them.
> **Level 1 (1–3 marks)**: *For simple description; the more detail, the more marks.*
> **Level 2 (4–5 marks)**: *Relating your description to the other features on the map raises your answer to this higher level.*

2 Choose one of the market gardening areas shown in Figure B other than the Fylde or another area you have studied. Draw an annotated sketch map, similar to the one in Figure C, to show the physical and human factors affecting it. (10)

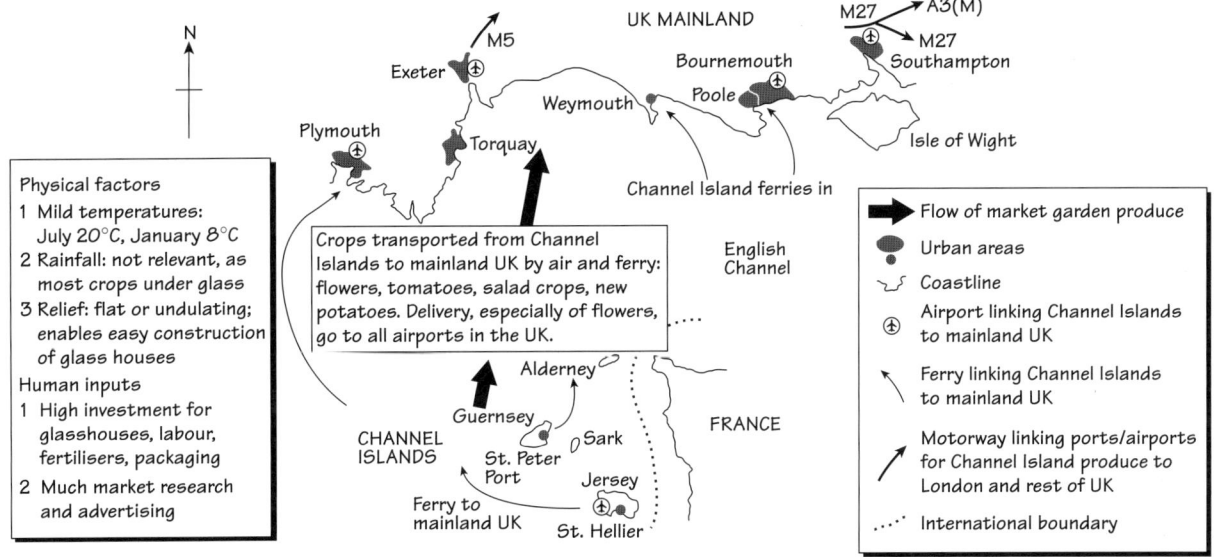

D A model answer using the example of the Channel Islands.

Improve Your Mark!

The Channel Islands have been chosen to demonstrate a response to this question to provide you with another short case study. Marks are not really given for the actual sketch map that you draw, but for the details in the labels or annotations you put on it. Annotations are a good way to express a considerable amount of information in an efficient manner. Your map must include details of physical and human/economic input. Study the labels here, which show you examples of these inputs and give information on how the produce is transported to mainland UK, so that you can develop your own skills. You can practise doing the same, of course, with any case study you choose. (Good quality annotations are a higher-level skill if used in your coursework and could potentially give you top marks in the data presentation section).

Level 1 (1–4 marks): *Would be given for a simple map with few labels.*

Level 2 (5–7 marks): *Awarded for greater detail, although the answer might not consider all possible points.*

Level 3 (8–10 marks): *A full answer considering physical and human factors is necessary here.*

3 Imagine that you wish to set up a market gardening firm. Suggest four factors that you would consider about a particular location before setting up there.　　(4)

Four factors which need to be taken into account when setting up a market gardening business would be:

- the availability of money to provide enough investment for polytunnels, watering systems, high quality seed, fertilisers, etc.

- the availability of labour, some of it skilled

- access to markets; the costs of transport must not be too high and suitable means of transport must be available, for example in the case of the Channel Islands, there are ferries and air services, both of which carry produce daily

- physical factors, such as hours of sunshine for ripening.

Improve Your Mark!

Note that human factors are picked out much more than physical ones here because the technology of market gardening can reduce many potential climatic problems. Location with regard to markets is much more important!

Level 1 (1–2 marks): *Would be given to an answer with fewer than the required four factors or where little justification was offered.*

Level 2 (3–4 marks): *For full marks four reasons, each with some explanation, must be given.*

CHAPTER 8 deals with the subject of resources, which is covered in six sections. The first sections include: what are resources?, non-renewable energy resources, renewable energy resources, world energy and appropriate technology. The final section is a detailed case study of the Alaskan oil industry. Other studies in the chapter include the Itaipù dam in Brazil and the Wairakei geothermal power station in New Zealand.

What are Resources?   118

After defining resources as features of the environment that are needed and used by people, this section introduces the division of resources into non-renewable and renewable groups. Renewable resources are then further subdivided into flow resources such as water and sustainable resources such as fish or trees. It is important that pupils understand this subdivision and that sustainable resources are only sustainable through careful management – for example, overfishing or over-logging can degrade the resource. The increasing demand for resources is explained and then the importance of sustainable development is reiterated.

Energy Resources   119

The division of energy resources into renewable and non-renewable groups is clearly set out. The term *fossil fuel* is introduced, but not fully defined. Here is a dictionary definition: 'a hydrocarbon deposit that consists of the remains of animal or vegetable life from past geologic ages and that is now in a combustible form which is suitable for use as fuel; for example, oil, coal, or natural gas'.

Non-renewable Energy Resources   120 – 121

This section focuses on brief studies of the advantages and disadvantages of coal, oil, natural gas, nuclear energy and fuelwood. Figure 8.6 on page 121 is a very useful summary of the arguments for and against nuclear power.

Activity Sheet 8.1 studies change in the UK coal industry, testing the pupils' skills in reading and drawing graphs, and in explaining the changes affecting the industry.

Activity Sheet 8.2 studies the UK oil industry, providing useful production statistics.

Activity Sheet 8.3 is at a high level. It requires pupils to draw divided bars and pie graphs on copies of the map of Europe on page 9 of the *Teachers's Resource Book*. Nuclear energy might be an important energy source for the future because it can produce large amounts of electricity with little pollution and no contribution to global warming. Of course, the development of nuclear fusion would result in a virtually limitless source of energy.

Renewable Energy Resources   122 – 125

A study of the Itaipù dam in Brazil highlights the advantages and disadvantages of hydro-electric power. The North Island of New Zealand is the setting for the case study of the Wairakei geothermal power station. Examples of the use of geothermal energy in the UK can be found at the International Geothermal Association's website, www.iga.igg.cnr.it. Charlestown Junior & Infants School in St. Austell, Cornwall, is now heated and cooled using geothermal energy. The school's energy system absorbs heat from the Earth using renewable energy collectors installed in ten boreholes drilled in the school grounds and transfers it to the school using high-efficiency heat pumps. A second scheme at Southampton is a district heating project using water at 76°C from a depth of approximately 1800 m below the city. The project, launched in 1987, provides heat to the new Western Esplanade district in the centre of Southampton, including a shopping centre, offices, hotel and central baths. Geothermal energy provides 87% of the heat requirements and covers daily fluctuations in demand by means of storage tanks. A coal-fired boiler provides back-up.

Wind, solar, hydrogen, tidal, waves and biomass energy sources are also studied.

Activity Sheet 8.4 provides details of landfill gas, which had grown rapidly to become the most important form of renewable energy in the UK by 2001.

Activity Sheet 8.5 is focused on wind power and includes an Internet-based activity. Information is provided on the largest wind farm in the UK at Bowbeat Hill in Scotland. There are also questions on solar, tidal and wave power.

World Energy   126 – 127

The two maps (Figures 8.23 and 8.24 on page 126) showing energy production and energy consumption per capita deserve careful study. Ask pupils to consider the implications of the patterns revealed by the maps in terms of world trade in energy, levels of economic development and geo-politics. Use Figure 8.25 on page 127 to draw bar graphs of consumption per capita in the USA, UK, Japan, Malaysia, Brazil, Egypt, Kenya and Rwanda – this will make the contrasts very clear to the pupils. The very high energy use in the USA, its causes, effects and possible consequences can be discussed.

Case Study 8 Energy and the Environment – Oil in Alaska 📖 128 – 131

Alaska provides a useful case study of the search for oil moving to increasingly remote areas of the earth. The environmental issues resulting are studied in detail.

Case Study Extra Coal Mining in South Wales

South Wales provides a study of the growth and decline of the coal mining industry. Opencast mining is featured, as well as the effects on communities of the closure of coal mines. It is useful to refer to pages 138 and 139 in the pupil book, which include a case study of the iron and steel industry in South Wales.

The last question posed in this section takes an example from County Durham, where a mining village was actually swept away following the closure of the mine, and provides a fascinating contrast with the experience in South Wales, where communities have hung on, in some cases successfully, in others less so, but in all cases with great difficulty. The moral and political implications of the different approaches deserve discussion.

1 Study Figure A.

   a) How many mineworkers were there in (i) 1960, (ii) 1980 and (iii) 2000? (3)

   b) How many mineworkers' jobs were lost between 1960 and 2000? (1)

   c) During which five-year period did the greatest decline in coal output occur? (1)

   d) Compare the changes in deep-mined and opencast coal output. (4)

(i)

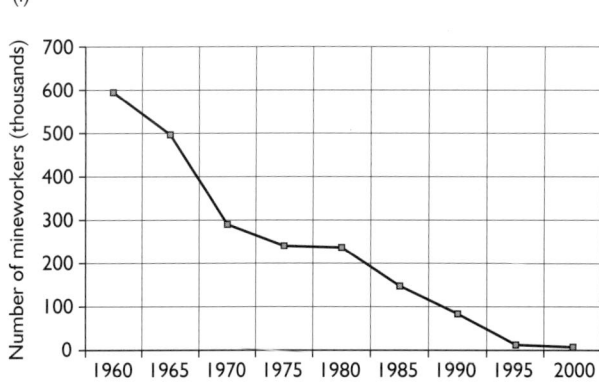

(ii)

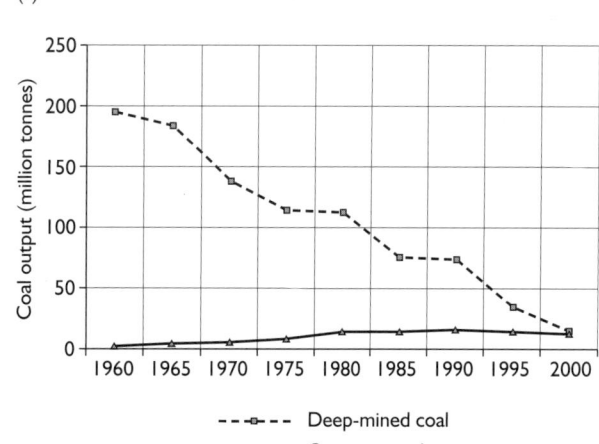

A Change in the British coal industry, 1960–2000.

2 a) Draw a line graph to illustrate the statistics given in Figure B. (5)

   b) Describe what happened to the number of deep coal mines between 1960 and 2000. (2)

   c) Why do you think this change has taken place? (2)

Year	1960	1965	1970	1975	1980	1985	1990	1995	2000
Number of deep mines	710	535	310	245	223	145	81	31	19

B The number of deep coal mines in the UK, 1960–2000.

3 Study Figure C. Describe and explain the changes which took place in the sources of energy consumed in the UK between 1970 and 2001. (6)

4 a) Briefly describe the three changes affecting the British coal industry revealed by the statistics in the graphs. (3)

   b) Give three reasons to explain the changes you have described above. (6)

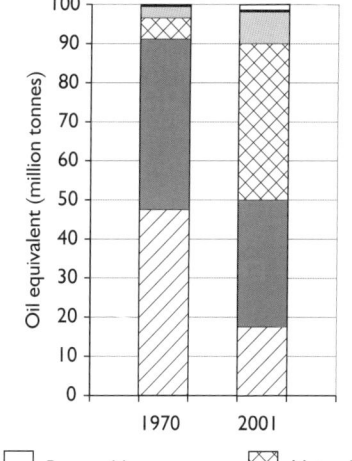

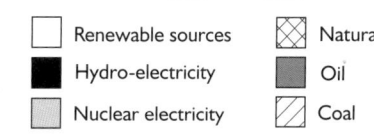

C Sources of energy consumed in the UK in 1970 and 2001.

1 Study Figure A.

a) Describe the changing pattern of oil production in the UK between 1970 and 2000. (3)

b) What area of oil production did the UK develop after 1975? (1)

c) What factors may explain the fall in production of oil in 1990? (3)

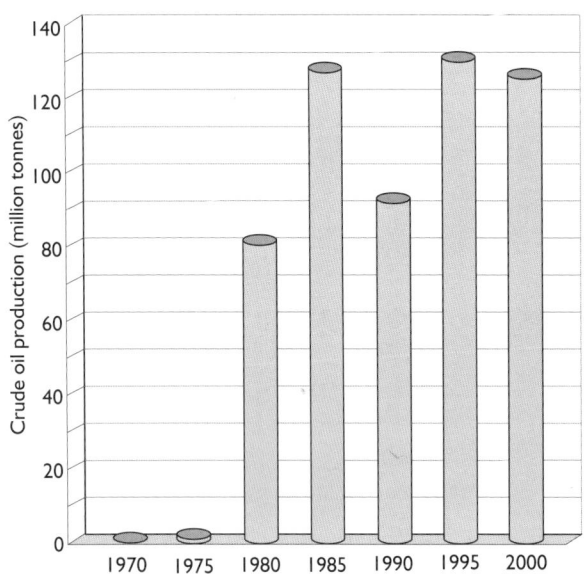

A Oil production in the UK.

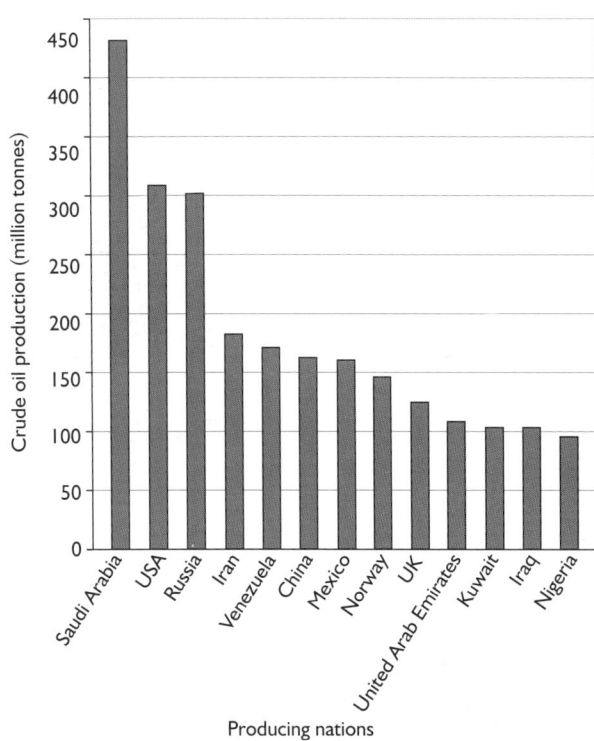

B Major crude oil producing nations.

2 Study Figure B.

a) What is the total oil production of (i) Saudi Arabia and (ii) Mexico? (2)

b) The thirteen nations shown in Figure B produce a total of 2400 million tonnes of oil out of a total world production of 3300 million tonnes. What percentage of the total world production is produced by (i) the thirteen nations combined, (ii) Saudi Arabia and (iii) Mexico? (3)

c) (i) Name the five nations in the Middle East included in the top thirteen nations. (5)

(ii) What is the combined oil production of these five nations? (2)

(iii) What percentage of the world total is this? (2)

3 Give three advantages and three disadvantages of oil as a source of energy. (6)

4 In the 1970s there were fears that the world's oil would run out by the early years of the 21st century, but this did not happen. Why not? (4)

1 How does a nuclear power station generate electricity? (3)

2 Study Figure A.

 a) Draw a divided bar graph to show the percentage of total electricity that is generated by nuclear power stations in the ten countries with the highest percentages. (5)

 b) On an outline map of Europe, draw small pie graphs to show the percentage of total electricity that is generated by nuclear power stations in each European country listed in the table. (7)

 c) Give two reasons why in France, Belgium and South Korea high percentages of the total electricity is produced by nuclear power. (2)

Country	% of total	Country	% of total	Country	% of total
Argentina	14	Hungary	43	Slovenia	43
Belgium	59	Japan	31	South Korea	40
Bulgaria	37	India	2	Spain	36
China	1	Lithuania	87	Sweden	42
Czech Republic	29	Mexico	3	Switzerland	38
Finland	32	Netherlands	5	UK	26
France	78	Russia	13	Ukraine	33
Germany	30	Slovakia	54	USA	21

A The percentage of total electricity produced by nuclear power in selected countries.

3 Study Figure B.

 a) Why might nuclear energy be an important energy source for the future? (2)

 b) Continue the conversation between the people in Figure B, making two more points for each. (6)

I understand your fears, but I'm afraid to say that they are the result of your ignorance. Future electricity supply has limited options. Coal, oil and gas all pollute the atmosphere and contribute to global warming. Renewable sources of energy can never be major contributors to electricity production – the best hydro-electric sites are already in use or in areas too beautiful to be flooded, the wind does not blow all the time and the sun does not shine at night. Future electricity has to depend upon nuclear power.

Nuclear power should be stopped. There have been too many accidents such as Chernobyl that have allowed radioactive products to escape. Nuclear power stations can catch fire through accident or the work of terrorists. The Irish Sea is increasingly contaminated by waste from the Sellafield reprocessing plant in Cumbria. Renewable forms of energy can offer safer sources of power.

However safe you say nuclear power is now, the radioactive waste you produce will remain a threat for ten thousand years…

B The nuclear power debate.

1 Study Figure A.

  a) Which is the most important form of renewable energy in the UK? *(1)*

  b) What is meant by the terms *solar photovoltaics, municipal solid waste combustion* and *farm waste digestion?* *(3)*

2 Study Figure B.

  a) What is landfill gas? *(1)*

  b) What was the total landfill gas generation capacity in 2001? *(1)*

  c) How is landfill gas used to generate electricity? *(3)*

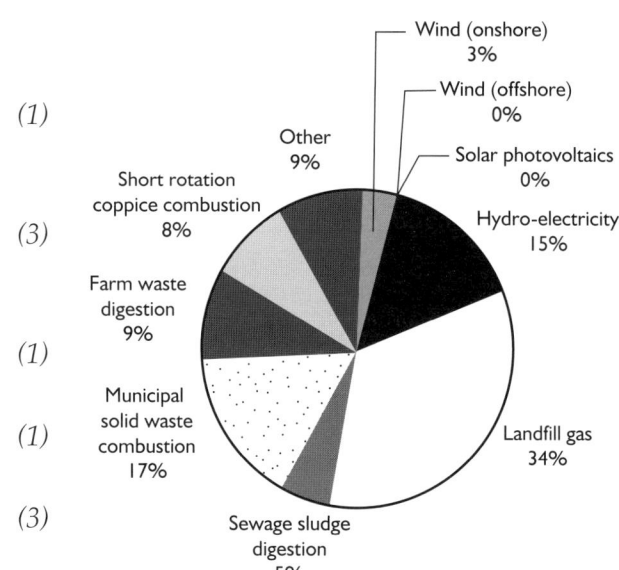

A Renewable sources used to generate electricity in the UK, 2001.

What is landfill gas?

Landfill gas is the result of the decomposition of organic waste within landfill waste sites. Landfill gas is Britain's leading form of renewable electricity generation. The first projects began in 1990 and by 2001 total landfill gas generation capacity was 418 MW (megawatts). There are over 300 projects; the largest has a capacity of 18 MW, although the vast majority are in the range of 1–3 MW.

How a landfill gas site generates electricity

Once the landfill pit has been filled, it is capped. Pipes called wells are sunk through the cap into the waste where they can access the gas. The wells run to a manifold where they are joined to a larger pipe called a main, which carries the gas to the electricity generation plant. The gas is connected to a pump, which pulls gas through the network. Having passed through filters to remove any particles and a cyclone to remove moisture, the gas is pumped into the engine, which drives an alternator. The alternator creates the electrical current, which is supplied to a transformer. The transformer increases the voltage and supplies electricity directly to the national grid. Any excess gas is burnt off at a chimney called a flare stack.

B How a landfill gas site generates electricity.

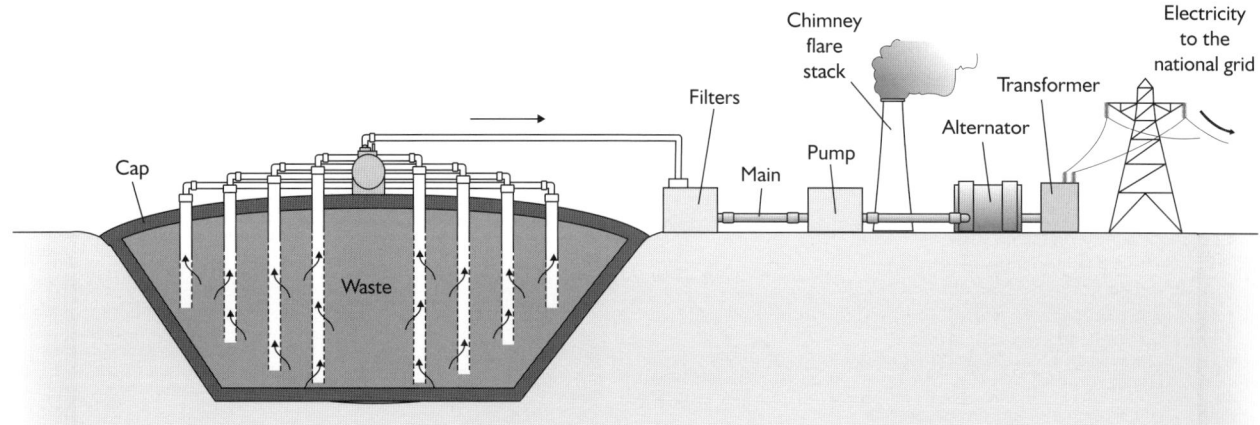

C Electricity production from landfill gas.

3 a) Using Figures B and C on page 90 how does a hydro-electric power station generate electricity? (3)

   b) Give the location of a major hydro-electric power scheme. (1)

   c) Give four factors which are involved in the location of the power scheme you have named. (4)

   d) What advantages and disadvantages does this hydro-electric scheme have? (6)

4 a) What is meant by *geothermal energy*? (1)

   b) Study Figure 8.14 on page 123 of the pupil book. Describe how the hot, dry rock method of generating electricity works. (3)

   c) Name a country which makes use of geothermal energy and give two ways in which the energy is used. (3)

1 Study the extract below.

   a) Where is Britain's largest wind farm? *(1)*

   b) (i) What is the generating capacity of this wind farm? *(1)*

      (ii) How does this compare with the generating capacity of coal and oil-fired power stations? *(1)*

   c) Describe the feelings of the local councillors about the Bowbeat Hill Wind Farm. *(4)*

   d) What are the environmental advantages and disadvantages of wind farms? *(8)*

Bowbeat Hill Wind Farm

Britain's largest wind farm is located on moorland in the Moorfoot Hills, some 7.5 km north-east of Peebles in south-east Scotland. The Bowbeat Hill Wind Farm has twenty-four 1.3 MW (megawatt) turbines. The towers are tubular, made of steel and painted light grey. The blades are made of glass-fibre reinforced polyester. They are light grey because this is the colour that is most inconspicuous. The 76 m high turbines (46 m steel towers, with 60 m diameter blades) have a total generating capacity of 32 MW – enough electricity to cater for the annual power requirements of around 22 500 homes (by comparison, large oil and coal-fired power stations can generate up to 2000 MW each). The scheme's annual output of electricity will replace enough fossil fuel generation to prevent the emission of around 50 000 tonnes of the greenhouse gas carbon dioxide (CO_2) a year.

The wind turbines start operating at wind speeds of around 15 km per hour and reach their maximum power output at around 55 km per hour. At very high wind speeds, i.e. gale force winds (80+ km per hour), the turbine blades are feathered and shut down to prevent racing away.

DESPITE previous concerns over its impact on the surrounding countryside, Bowbeat Hill Wind Farm, near Peebles, has begun generating electricity for the first time. Built by Powergen Renewables, the wind farm had raised eyebrows among leading members of both the Peebles and Innerleithen communities, who felt it would be visually detrimental to the area.

Chairman of Peebles Community Council, John Moore, explained his feelings on the matter of wind farms to the Peeblesshire News. He said: 'We at the Community Council discussed plans for the wind farm and as far as we were concerned we had reservations about it. I had seen similar ones in Cornwall and Denmark and concluded that they were not the most attractive things ever built. However, we have reached a stage now where the benefits outweigh this. It is not just a case of the odd windmill being put up by a crank — these systems seem to be the way forward.'

Innerleithen Councillor Reid Meikle is also in favour of the project. He added: 'Although these turbines resemble ghost-like figures on the skyline, I am in support of them. They are stuck up in the hills in the middle of nowhere and apart from the walking fraternity, no-one is likely to be anywhere near them. We have to move with the times and if wind farms like this one are able to provide local houses with environment-friendly power, then that can only be a good thing.'

*From **Peeblesshire News** 26 September 2002.*

---

2 Visit the British Wind Energy website at www.bwea.com to find out the latest news about wind turbines and to see a map of the wind farms in the UK. Where is the nearest wind farm to your home? *(1)*

3 a) Why is solar energy unlikely to ever be an important energy source in the UK? *(2)*

   b) In which parts of the world might solar energy be most important? *(2)*

---

4 Study page 125 of the pupil book.

   a) Name the location of a tidal power station. *(1)*

   b) How does a tidal power station generate electricity? *(3)*

   c) Link the statements A–E with boxes 1–5 on Figure A (see Figure 8.20 in the pupil book). *(5)*

     A Tide directed into a set of tunnels, each of which has a turbine.

     B Road built across dam reduces driving times and costs (30 km shorter).

     C Locks for ships.

     D As tide recedes the blades of the turbines reverse.

     E Incoming tides (twice daily) have a range up to 11.6 m and can reach 20 km/hr; maximum at spring tides, but no seasonal variation.

   d) Name two possible disadvantages of large-scale tidal power projects. *(2)*

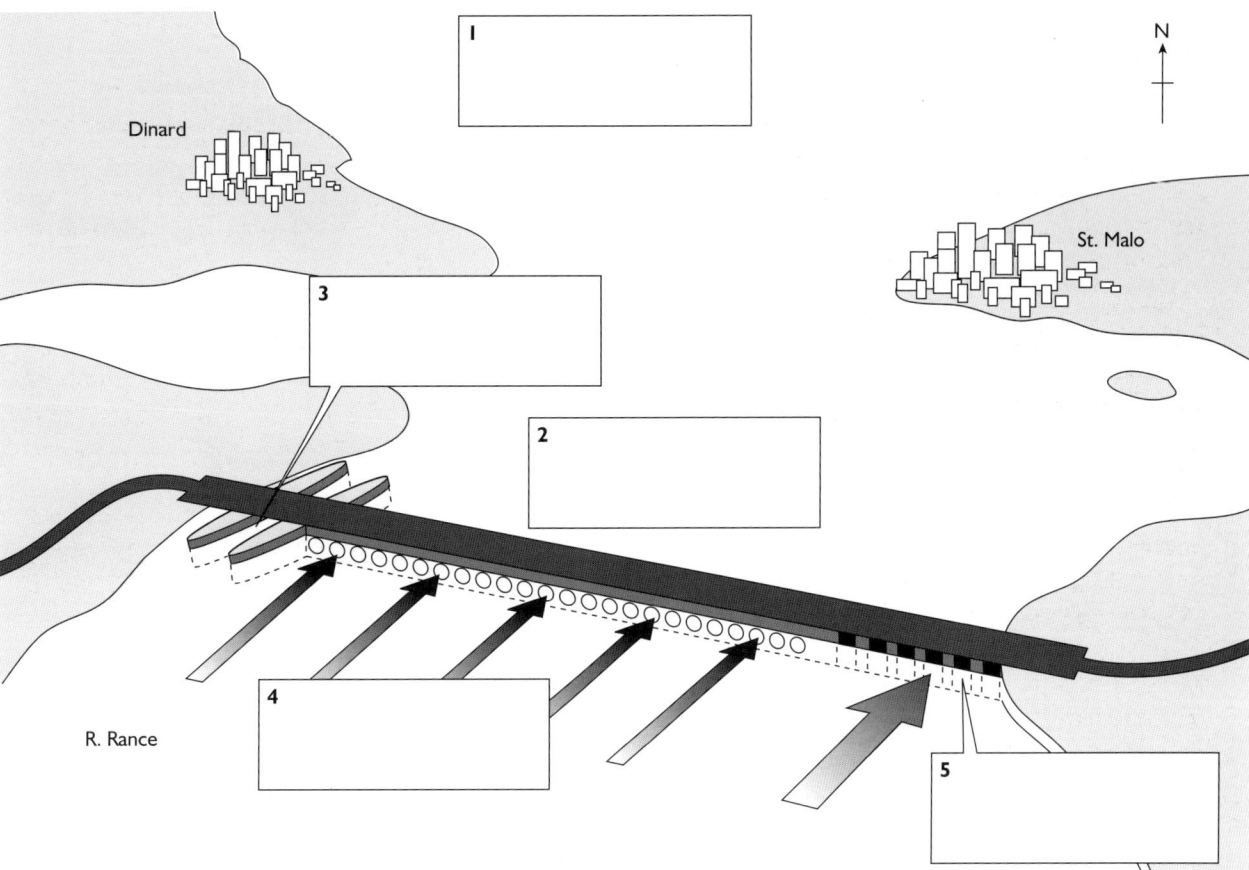

A The Rance Tidal barrage in France.

5 Draw a labelled diagram to show how a wave energy collector operates (see Figure 8.21 in the pupil book). *(6)*

Coal Mining in South Wales

The Historical Legacy

Before 1800 South Wales was a sparsely populated farming region. The only industries were small-scale iron smelting and coal mining. Drift mines and water-powered ironworks using charcoal were sited in the valleys of the coalfield. With the replacement of charcoal by coke and of the waterwheel by the steam engine, coal became a vital industrial fuel.

South Wales, with its extensive coal resources, rapidly became one of Britain's greatest industrial areas (see Figure A).

The population of South Wales increased rapidly. People flocked to the valleys of the coalfield area, especially from mid-Wales, Shropshire and south-west England, in search of work in the coal mines, ironworks and other factories. New towns grew as the valleys began to fill up with mines, ironworks, canals, roads, railways, chapels and row upon row of terraced houses (see Figure B below and on page 95). The population of the Rhondda Valley rose

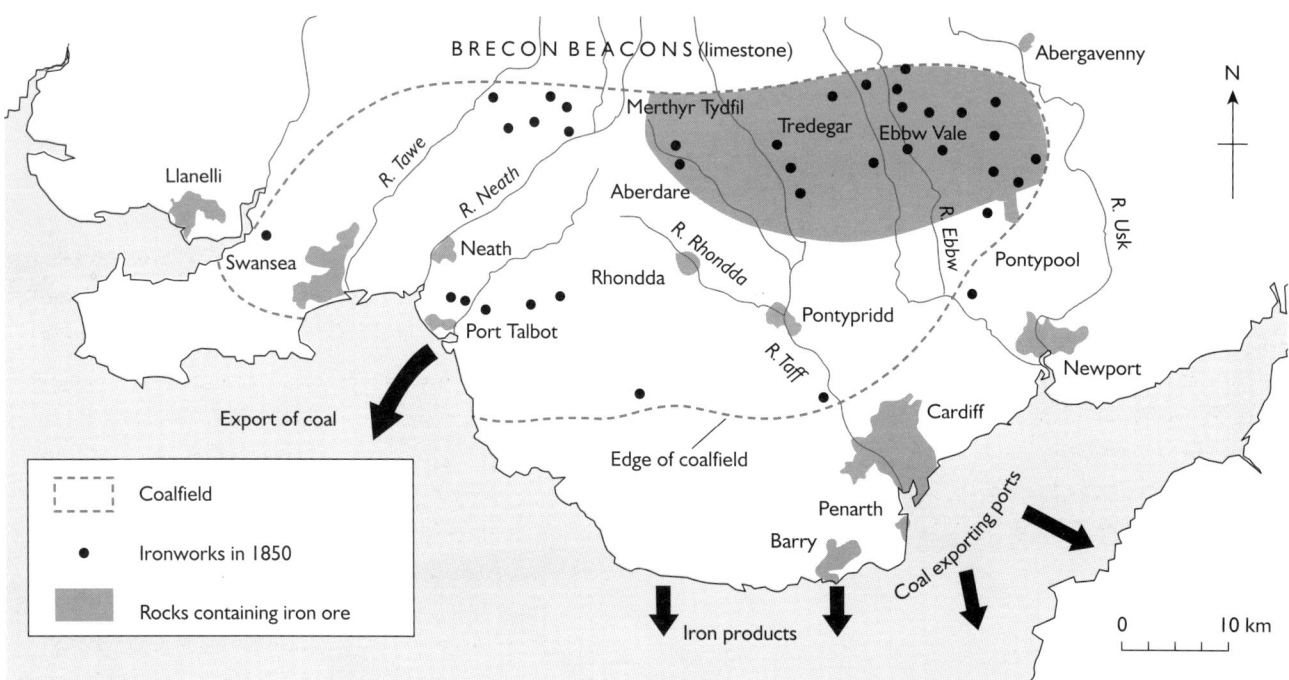

A South Wales.

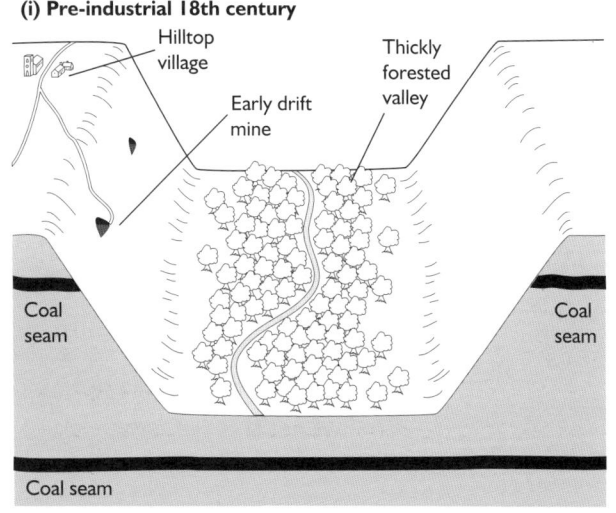

(i) Pre-industrial 18th century

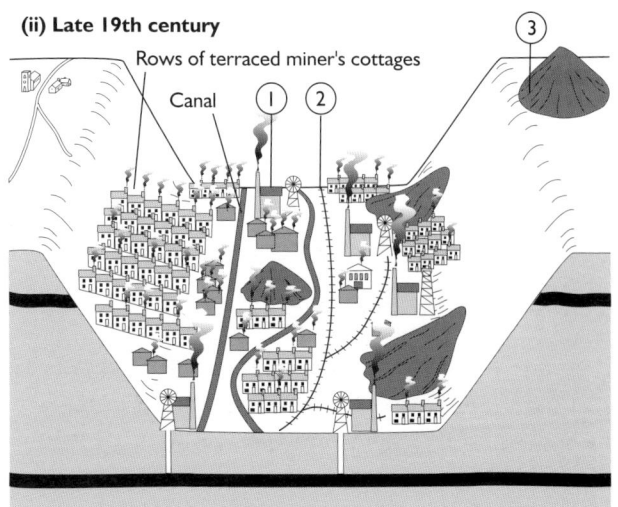

(ii) Late 19th century

B Change in a South Wales valley (continues on page 95).

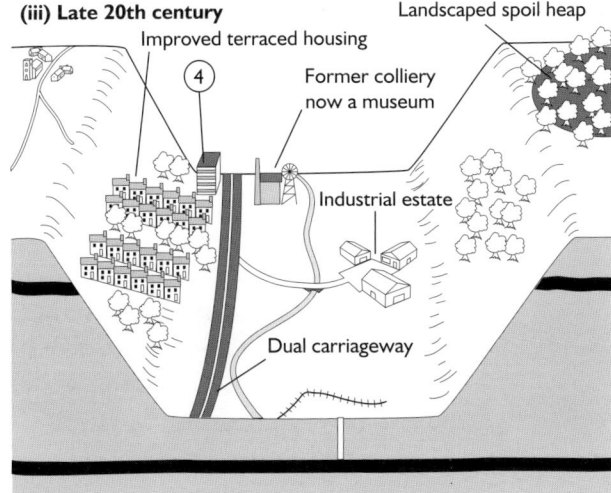

(iii) **Late 20th century**
Improved terraced housing
Landscaped spoil heap
4
Former colliery now a museum
Industrial estate
Dual carriageway

B Change in a South Wales valley (continued).

from under a thousand in 1801 to 140 000 by 1921. Living conditions were often poor, with up to fifteen people crammed into a tiny terraced house.

Employment in the coal mines of South Wales rose to a peak of 270 000 jobs by 1913 when there were 613 mines in operation. However, over-dependence upon coal and steel led to disaster in the late 1920s when the Great Depression reduced demand. Coal output dropped from 46 million tonnes in 1920 to 28 million tonnes in 1936. In the same period, over 300 000 people moved away from the valleys in search of work, mainly to south-east England and the Midlands (see Figure C).

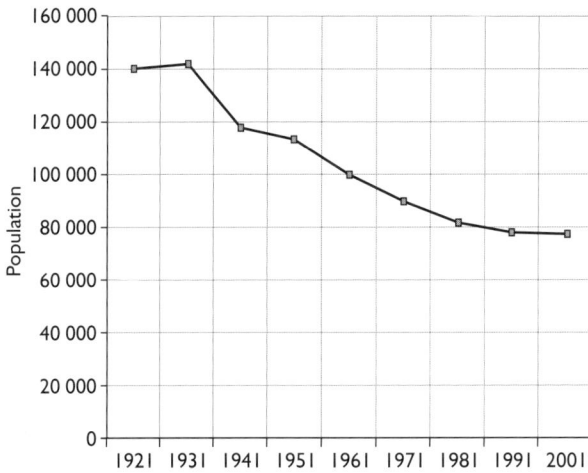

C Population of the Rhondda Valley.

A gradual decline in demand for coal meant that the slow decline of mining in South Wales continued throughout the 20th century until by 1990 only five mines remained open, employing under 10 000 people. Within four further years,

only one mine, the Tower Colliery, remained and it was bought by its own miners with their redundancy payments. Taken together with a second operating colliery, the Betws drift mine, total deep mining employment in South Wales amounted to about five hundred jobs in 2003.

Opencast Mining

There were eleven opencast coal mines in South Wales, mainly in the Neath and Port Talbot areas in 2002. They produce 80% of South Wales' coal production. They are quarry sites, producing coal by stripping away the cover of rock and soil (the overburden) to expose the coal seam (see Figure D on page 96). Opencast mines have been criticised on several environmental issues:

■ loss of farmland

■ loss of hedges and other wildlife habitats

■ noise and vibration from heavy machinery, such as vast dragline excavators and mechanical shovels

■ coal dust blowing from the site onto nearby land and homes causes nuisance and has also been linked to breathing diseases in children

■ increased traffic with the associated danger and pollution on local roads.

Pressure groups have protested against plans for new opencast mines. Protests turned to violence in 1996 when protestors were forcibly evicted from makeshift camps at the Selar opencast site owned by Celtic Energy Ltd. Although some opencast operators have been fined by the Environment Agency for polluting rivers, other operators have taken several measures to reduce the environmental nuisance:

■ spoil material has been built up into sound baffle mounds encircling the sites

■ water spraying helps to reduce the amount of windblown dust

■ the sites are restored to beneficial uses such as freshwater lakes, nature reserves, sports fields and golf courses after use.

In the UK as a whole there were 55 opencast mines in 2002, producing 14 million tonnes of coal per year (compared with 17 million tonnes of deep-mined coal).

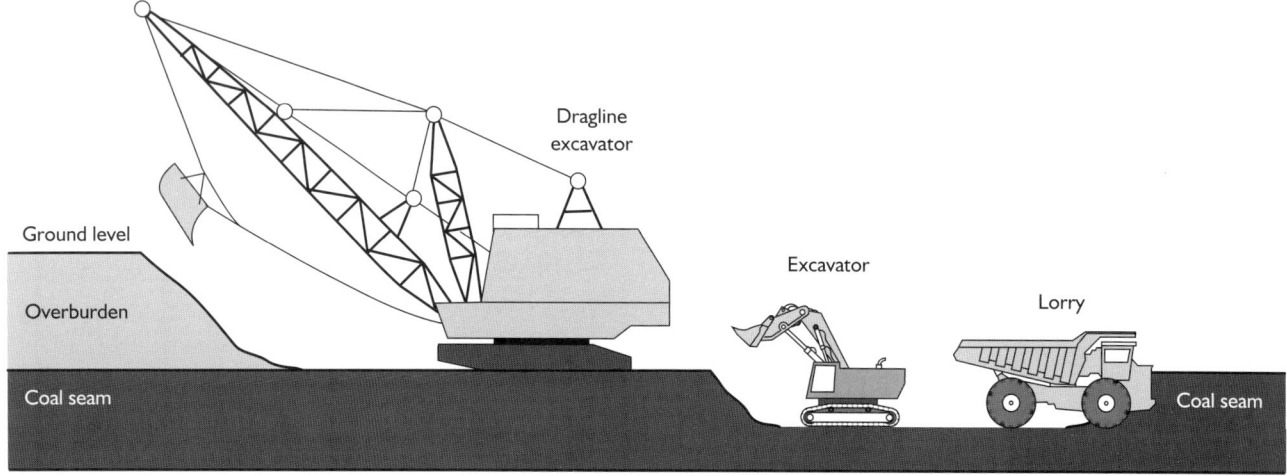

D Opencast mining.

The Effects on Communities of the Closure of Coal Mines

When coal mines close a 'reverse multiplier effect' (or 'spiral of decline') operates on the local economy. Unless effective measures are swiftly put in place by the government and local authorities, a process of gradual de-industrialisation and depopulation starts.

- For every miner's job that is lost as the result of the closure of a colliery, further jobs are lost elsewhere in the area. The most immediate impact is on the industry's suppliers, including engineering and haulage firms, and contractors.

- Local shops and other consumer services also lose business because of loss of income in the community even if, in the short term, the blow is eased by redundancy payments.

- In the longer term, areas that lose jobs suffer from out-migration and as people leave to find work elsewhere, they add a further downward twist to local spending and the health of the local economy.

- Eventually, if the population falls enough, even employment in public services, such as schools and hospitals, begins to decline.

The closure of the collieries also impacts on family life. The miners lose their focus in life and remain at home. The lack of skills for alternative employment means that the search for a new job is unsuccessful for many. The prospect of poorly paid, low-skilled jobs or of never working again causes depression in some miners. Family life is strained by the tensions created by unemployment. As people move away, the community spirit and life of the mining towns and villages also ebbs away. This in turn creates a range of social problems, including conflict within families, separation and divorce, and an increase in drug use, crime and vandalism, especially amongst the young.

By no means all ex-miners fall into despair nor do all ex-mining communities fare badly. Some miners have found new jobs that are well paid, some have become successfully self-employed, others spent their redundancy payments on establishing successful businesses and some have found contentment in remaining unemployed and following hobbies and pastimes such as gardening.

Regenerating the Welsh Valleys

Since the 1930s, the valleys have been helped by government regional assistance schemes, but with limited success. Although some large factories have been built in the region, such as Hoover washing machines at Merthyr Tydfil, the valleys as a whole have few attractions for large-scale industry. There is still some derelict land, but there are few flat sites for large factories and the area is relatively inaccessible (the M4 motorway passes well to the south). Much of the new industrial development in the valleys has been concentrated in the most favoured parts of the region; the modern A465 Heads of the Valleys road has attracted the development of a number of industrial estates, such as the one at Hirwaun.

One positive aspect of the decline of the coal industry has been the growth in heritage tourism which it has allowed. A number of mines have re-opened as museums. The Ty Mawr/Lewis Merthyr Colliery at Pontypridd, which closed in 1983, is now the centrepiece of a Heritage Centre which has recreated a mining village. The centre employs 250 people and attracts over a quarter of a million visitors each year. The Big Pit at Blaenavon, closed in 1980, has also been restored. Visitors can travel down the 100 m mine shaft and experience the coal miners' world.

Whilst the valleys have struggled, the coastal lowlands have prospered. As the site of the Welsh Assembly, there has been a growth in employment in government and administration in Cardiff. High-tech industry has developed along Severnside, including many foreign, especially US and Japanese, companies. Newport's Celtic Lakes Business Park has attracted several electronics and computer firms.

The economic focus of South Wales has moved from the coalfield valleys to the more accessible, milder lowlands, which have the flatter and more extensive sites required for modern industry. Parts of the valleys remain blighted by severe social and economic problems, relating to unemployment, long-term sickness and poverty.

Exam Practice Questions

1 Why did South Wales become an important industrial region? *(2)*

2 a) Figure B shows a model South Wales valley. Write suitable labels for the features numbered 1–4. *(4)*

b) Explain how a typical South Wales valley was altered by coal mining. *(4)*

3 a) How did the population of the Rhondda Valley change between 1801 and 1921? Why did this happen? *(3)*

b) Study Figure C. Describe and explain the pattern of population change in the Rhondda Valley between 1921 and 2001. *(4)*

c) What are the social effects resulting from the closure of coal mines in the valleys? *(4)*

4 a) Use Figure D to describe how coal is extracted from an opencast mine. *(2)*

b) What environmental problems are associated with opencast coal mines? *(3)*

c) What measures can mining companies take to reduce the environmental impact of opencast coal mining? *(3)*

5 There are alternatives to the policy of regenerating former mining areas adopted in South Wales. In the north-east of England, a policy of abandonment was followed. Leasingthorne (see Figure E on page 98) is a village in County Durham. The village developed as the result of the opening of a coal mine in 1842. By 1921, there were 1061 inhabitants in a thriving community, which included a school and several shops and other services. Leasingthorne Colliery closed in 1965 when the coal was worked out and the effect on the community was devastating. By 1971 the village had only 157 inhabitants and had lost all its shops and services. In 2002, the site of Leasingthorne had returned to open fields with only a handful of houses remaining.

a) Study Figure E. State three changes that took place at Leasingthorne Colliery between 1920 and 1961. *(3)*

b) Study the 1970 map. State six changes that took place in Leasingthorne during the 1960s. *(6)*

c) Why has the village of Leasingthorne disappeared? *(3)*

d) What are the advantages and disadvantages of a planning policy that aims at removing a whole community such as Leasingthorne? *(4)*

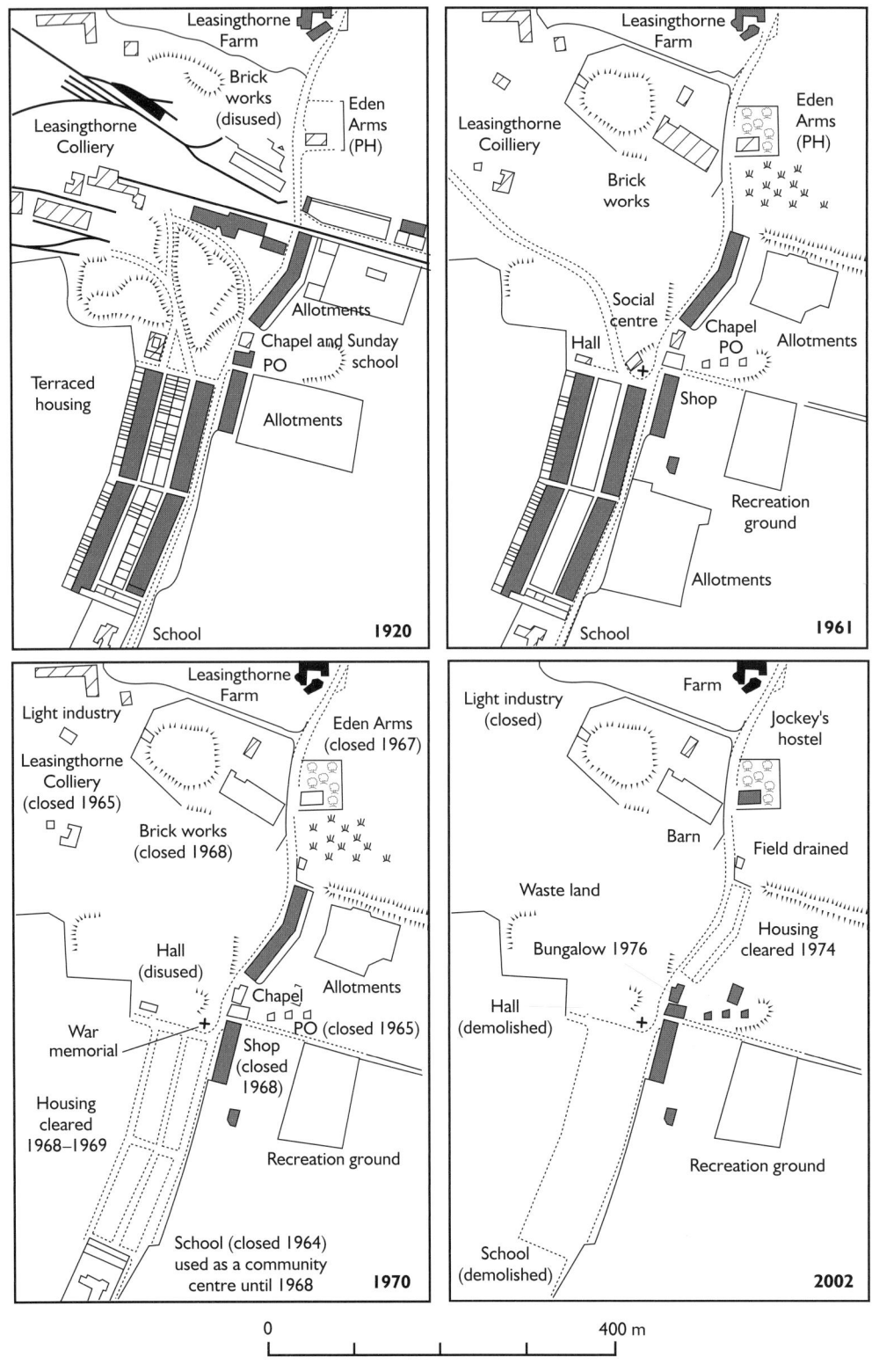

E Leasingthorne in 1920, 1961, 1970 and 2002.

Coal Mining in South Wales

These model answers are designed to show you how to get full marks for some of the longer questions in the Case Study Extra for this section. Read the question carefully and write your own answer. Then read the model answer and the examiner's notes to see what he or she is looking for when awarding the marks. Decide how many marks you think an examiner would give your answer. Decide how to change your answer to increase your marks. What will you do differently next time you answer a similar question?

5 a) Why has the village of Leasingthorne disappeared? (3)

> The village depended on the colliery as its major source of employment. When the colliery closed, followed by the brickworks, the village shops and services rapidly closed because people left the village, as shown by the clearance of many houses.

b) What are the advantages and disadvantages of a planning policy that aims at removing a whole community such as Leasingthorne? (4)

> The advantage of such a policy is that by removing the village following the closure of the colliery, Leasingthorne's main reason for existence, there is no need to attract new employment into the village – often an expensive process. Also services such as schools and post offices will not need to be kept open at a loss if the village is removed.
>
> The disadvantages of such a policy are mainly social. The community will be broken up and people will be forced to leave their homes and move away. The village may have acted as a service centre for surrounding farms and hamlets, which will be left more isolated.

<table>
<tr><td rowspan="3" style="writing-mode:vertical-lr">Improve Your Mark!</td><td>

Level 1 (1–2 marks): *Basic answer including reference to the effects on local people of the break up of their community.*

Level 2 (3–4 marks): *More detailed answer including reference to the break up of the community and the loss of services.*

Level 3 (5–6 marks): *Good, detailed, balanced answers including the advantages of not having to fund new employment and maintain loss-making services, and comments at Level 2 or above.*

</td></tr>
</table>

CHAPTER 9 deals with industry. This is covered under six sections: the industrial system and industrial location, changing locations, transnational corporations, the Pacific Rim, industry in LEDCs and sustainable development in LEDCs. An additional section investigates Osaka-Kobe and São Paulo as case studies of industry in a developed and a developing country.

Industrial Location 📖 136 – 139

A study of industrial location in the UK is followed up in Activity Sheet 9.1. The case study of the iron and steel industry in South Wales that follows makes a useful complement to the case study of coal mining in Chapter 8 of this *Teacher's Resource Book*.

Activity Sheet 9.1 provides further work on industry as a system and uses Figures 9.6 and 9.7 on page 138 to test pupils' understanding. It is a demanding sheet, well-suited to more able pupils.

High-Technology Industries 📖 140 – 141

The M4 corridor provides an example of an axis of high-tech industry. Business and science parks are illustrated by Tsukuba Science City in Japan. Question 6 on page 158 of the pupil book includes a useful map of the Cambridge Science Park.

Activity Sheet 9.2 picks up the M4 corridor case study and supplies details on the Windmill Hill Business Park in Swindon, which is illustrated in Figure 9.11 on page 140.

Government Policies in the UK 📖 142 – 143

A brief study of central government's regional development policies features enterprise zones and urban development corporations (UDCs). UDCs were wound up in 1998, although their impact continues today.

Transnational Corporations 📖 144 – 145

The sheer size and wealth of transnational corporations (TNCs) need to be stressed to pupils. They control 75% of world trade and over 50% of the world's manufacturing. Figure 9.19 on page 144 provides a very useful summary of the advantages and disadvantages of TNCs to a country. The example of the Ford Motor Company is used to show the global role of TNCs. The term *globalisation* has had a bad press in recent years, with anti-globalisation riots at major international conferences. The reasons for globalisation and its advantages and disadvantages are important points for discussion.

Activity Sheet 9.3 focuses on the advantages and disadvantages of TNCs and emphasises the importance of the development of communications technology in the growth of globalisation.

The Pacific Rim 📖 146 – 147

Using studies of Japan and Malaysia, these pages outline the reasons for the industrial success of the Pacific Rim. The concept of the newly industrialised country (NIC) is explained.

Activity Sheet 9.4 provides some useful additional statistics for Malaysia's trade and population.

Industry in LEDCs 📖 148 – 151

Starting with a detailed study of the difference between the formal and informal sectors, this section includes an eyewitness account of the informal sector in Rio de Janeiro. An additional idea is to ask pupils to give examples of informal sector employment they are aware of in their own area. The section then concludes with a study of appropriate technology in Kenya.

Activity Sheet 9.5 uses the examples of Bangladesh and Brazil to highlight the contrasts that exist between developing countries. Brazil has become a major industrial nation exporting aircraft and cars, while Bangladesh remains at a much earlier stage of industrial development. This is a demanding sheet, aimed at high ability pupils.

Case Studies 9a and 9b Industry in a City in a Developed Country: Osaka-Kobe and Industry in a City in a Developing Country: São Paulo 📖 152 – 155

These case studies highlight the contrasts between industry in cities in developed and developing countries. By choosing São Paulo, the similarities can also be highlighted because this city is the most industrialised of all the LEDCs. Some useful points on TNCs and globalisation can be made.

Case Study Extra Taiwan – A Newly Industrialised Country

Taiwan was one of the original four tigers of South-East Asia. This case study clarifies the path to economic development that many other developing countries are trying to follow. However, conditions were very different in the 1950s and 1960s, and it is a moot point whether other nations can successfully follow the four tigers. All had strong government, with martial law in the cases of Taiwan and South Korea, British colonial authority in Hong Kong and a single dominant leader in Singapore's fledgling democracy. Thailand, Malaysia and Indonesia have the best prospects for following Taiwan, but in each case the path will be a difficult one.

1 Figure A shows the steel industry as a system.

a) Decide which of the following are inputs, processes or outputs. Put each one in the appropriate box in Figure A. (3)

- Slag (waste)
- Smelting iron ore
- Limestone
- Cooling
- Steel
- Fuel

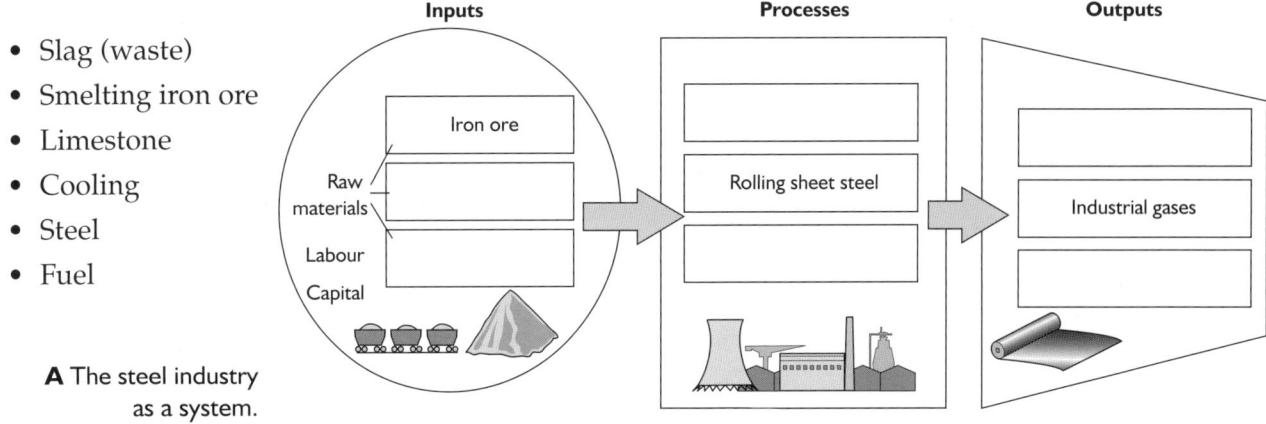

A The steel industry as a system.

b) Explain the meaning of the terms (i) *inputs* and (ii) *processes*. (4)

2 a) Figure B shows the traditional industrial areas of the UK (before 1970). Explain this pattern of location. (5)

b) Figure C shows the location of present-day industries in the UK.

(i) Explain why the pattern of location has changed from that shown in Figure B. (5)

(ii) Why have there been large-scale job losses in the UK's traditional industries? (3)

3 Describe and explain, using a regional example, how the location of the iron and steel industry in the UK has changed since the early 19th century. (10)

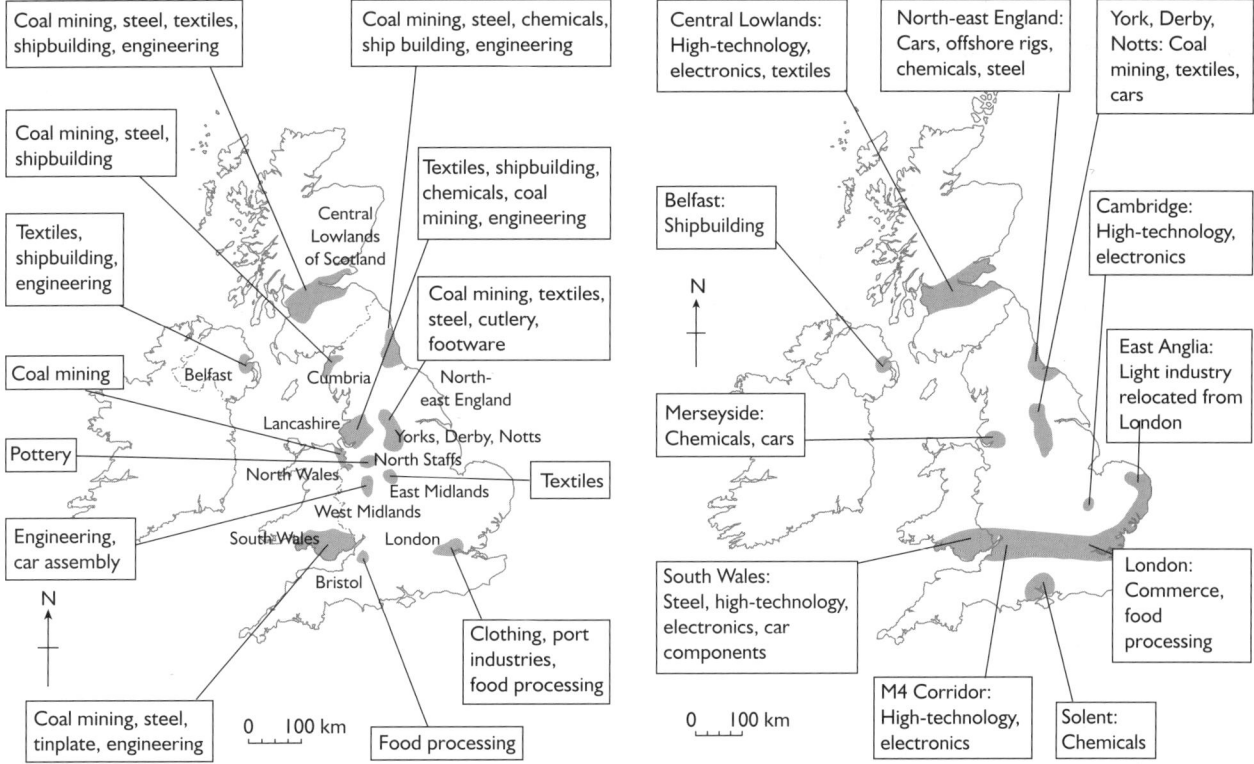

B Traditional industrial areas in the UK before 1970.

C Location of present-day industries in the UK.

1 Figure A shows the M4 corridor, one of the UK's major areas of high-tech industry.

   a) Define the term *high-tech industry*. (2)

   b) High-tech industry is said to be *footloose*. What is the meaning of this term? (2)

   c) Use your atlas to help you name the eight towns shown on Figure A by their first letters. (4)

   d) Why is the area shown on the map sometimes called 'the Sunrise Strip'? (2)

   e) List five advantages for high-tech industry of a location on the M4 corridor. (5)

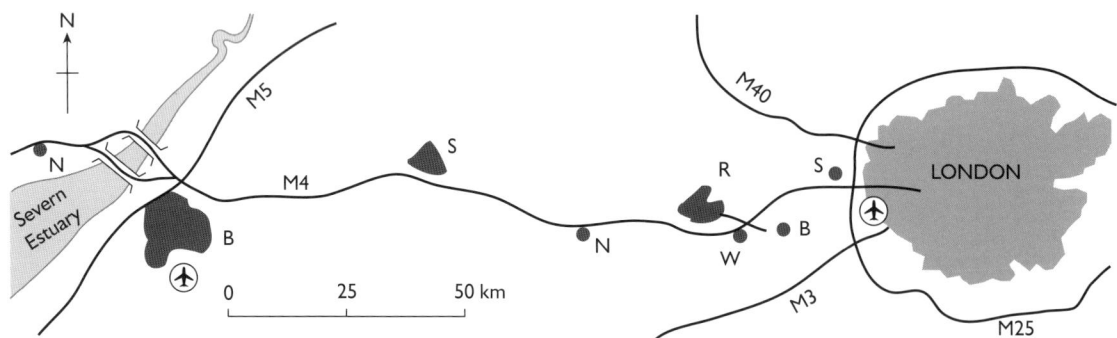

A The M4 Corridor.

2 Why do high-tech firms locate near to each other? (3)

3 The article below supplies some details on the Windmill Hill Business Park at Swindon, illustrated in Figure 9.11 on page 140 of the pupil book.

   a) What are the advantages of Windmill Hill as a location? (4)

   b) What has been done to make the environment of the business park attractive? (3)

   c) Why do you think the property developers went to such expense to make the environment attractive? (2)

   d) What type of industry has been attracted to Windmill Hill? (3)

Windmill Hill Business Park

Windmill Hill in Swindon is one of the country's most successful business parks with high-quality low-rise office buildings set among mature landscaping. Built in seven phases between 1984 and the present day on a greenfield site close to Junction 16 of the M4 motorway, the centrepiece of the business park is the 200-year-old windmill itself. It looks as if it has always been on this site, but in fact it was moved stone by stone from its original site on the Wiltshire Downs and rebuilt at Windmill Hill. This is a remarkable example of the lengths and expense to which developers are prepared to go to provide a distinctive and attractive working environment. The mature and varied landscaping at Windmill Hill is carefully maintained to create an overall sense of wellbeing and permanence, with pleasant tree-lined walks to help inspiration and provide relaxation.

All the buildings at Windmill Hill have a unique design. Each one is clad in reflective glass and has air conditioning, a highly flexible interior layout and raised floors for computer wiring.

Windmill Hill aims to provide an excellent location for company headquarters, research and development establishments and high-tech industry.

Current occupiers include:

• the national HQ of National Power

• the world HQ of Lucent Technologies Cable & Wireless

• PHH Vehicle Management

• McLean Homes South West

• Cendant Relocation UK

• Regus Business Centre

• Dialog Semiconductor

• VAT Clearing House.

---

1 a) What is a transnational corporation (TNC)? *(2)*

 b) Name three TNCs. *(3)*

2 Figure A shows the location of a TNC's factories and headquarters. The factories make electronic consumer goods.

 a) In which country is the headquarters of the corporation located? *(1)*

 b) Describe the distribution of the corporation's factories. *(4)*

 c) Name four countries where the corporation owns factories. *(2)*

 d) What are the benefits for the company of locating in several different parts of the world? *(5)*

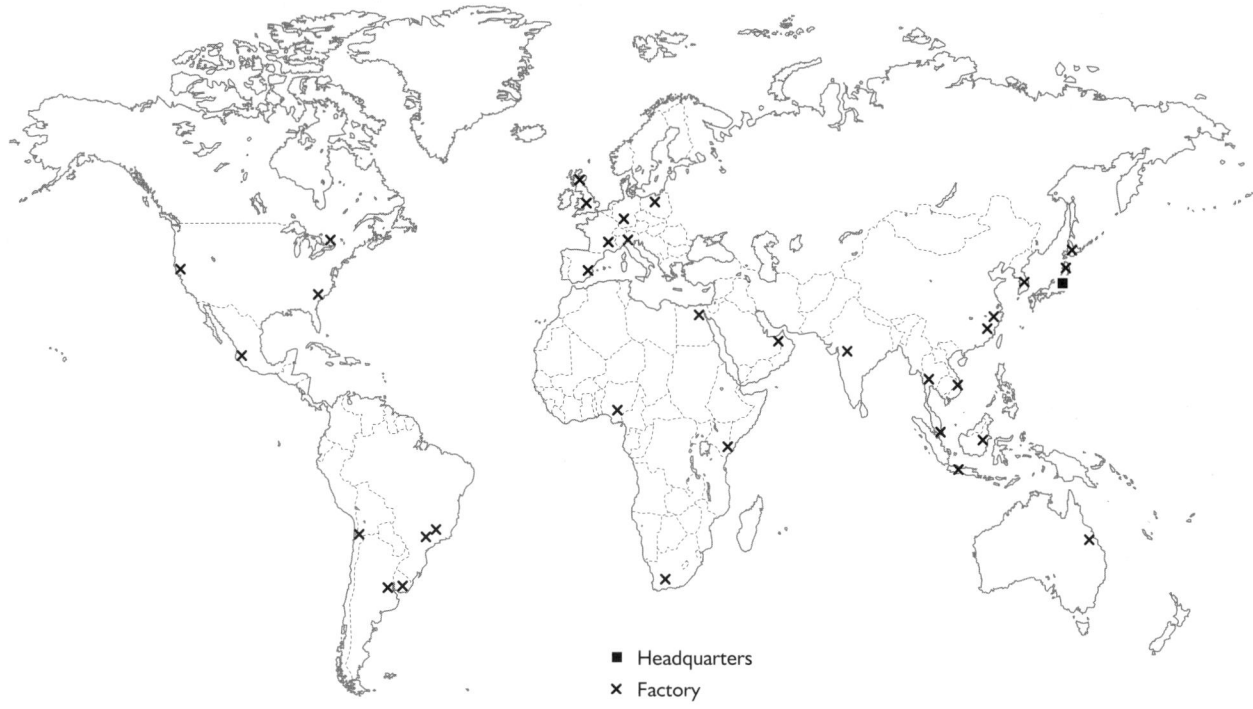

■ Headquarters
× Factory

A The locations of a TNC's factories and headquarters.

3 Give four advantages and four disadvantages of TNCs to the countries in which they own factories. *(8)*

4 Study Figure B.

 a) In what ways do TNCs dominate the world economy? *(3)*

 b) How has the development of communications technology assisted the growth of TNCs? *(4)*

5 Define the term *globalisation*. *(5)*

> Transnational corporations dominate the world's economy. For example, the 300 largest TNCs own or control more than one quarter of the whole world's productive assets. TNCs control about 70% of world trade.
>
> New communications technology allows TNCs to oversee and control their activities across the world. The internet gives managers instant communication wherever they may be. E-mail and video-conferencing mean that managers can cross and re-cross continents and oceans to control production without leaving their office desks.

B How the growth of communications technology has assisted TNCs.

1 What is an NIC? *(2)*

2 Study Figure A.

a) Describe the change in the export of (i) rubber, (ii) tin and (iii) manufactured goods from 1970 to 2002. *(3)*

b) (i) Explain why the changes you describe have taken place. *(4)*

    (ii) How have the lives of Malaysians changed as a result? Use Figure B to help you. *(3)*

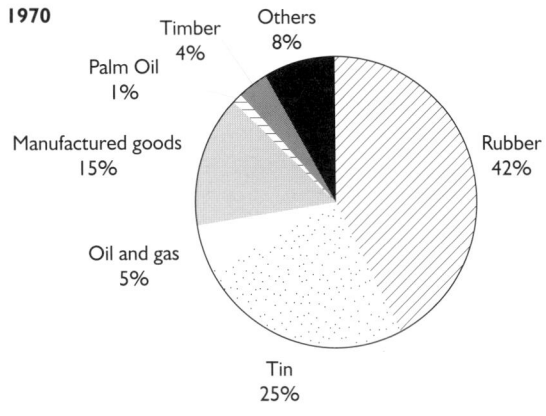

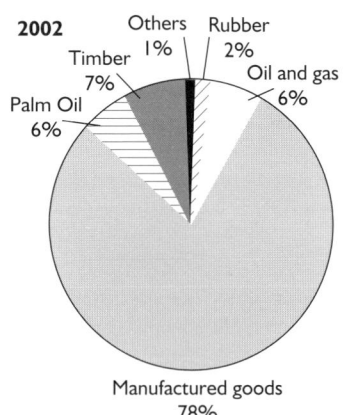

A Imports and exports from Malaysia in 1970 and 2002.

Year	1960	1965	1970	1975	1980	1985	1990	1995	2000
Population (millions)	8.1	9.4	10.9	11.9	13.8	15.5	17.9	19.8	23.2

Year	1960	1980	1995	2000
Crude birth rate (CBR)	44.0	30.9	26.8	25.6
Crude death rate (CDR)	15.0	5.3	4.7	4.6
Infant mortality rate (IMR)	86.0	23.9	10.3	7.7
Male life expectancy (years)	54.0	68.0	69.0	70.0
Female life expectancy (years)	56.0	72.0	74.0	75.0

B Population of Malaysia, 1960–2000.

3 a) Use the statistics in Figure B to complete the graph in Figure C. *(4)*

b) Describe the pattern of population change in Malaysia between 1960 and 2000. *(3)*

c) How does its growing economy help Malaysia cope with its rapid population growth? *(2)*

4 What problems do NICs such as Malaysia face as a result of their rapid economic growth? *(6)*

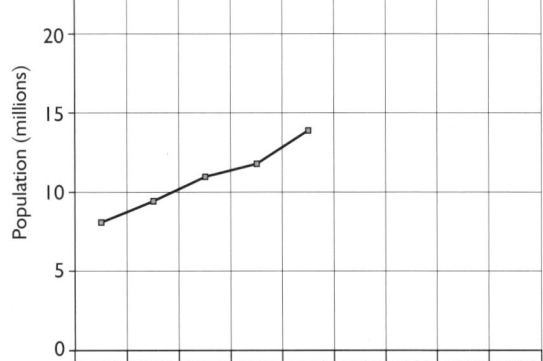

C Population of Malaysia, 1960–2000.

1 a) What is meant by (i) the *formal* and (ii) the *informal sector* of employment? (4)

b) Why do large numbers of people in LEDCs work in the informal sector? (2)

c) What risks are there in working in the informal rather than the formal sector? (3)

2 Study Figure A.

a) How does the manufacturing industry in Bangladesh compare with manufacturing in Brazil (also see pages 154–155 in the pupil book)? (5)

b) What reasons are there for the differences? (3)

Main industries in order of annual production		Main exports in order of value	
Bangladesh	Brazil	Bangladesh	Brazil
I Cotton textiles	I Textiles	I Clothing	I Electrical goods
2 Jute	2 Shoes	2 Jute and jute goods	2 Aircraft and components
3 Clothes	3 Chemicals	3 Leather	3 Motor vehicles
4 Tea	4 Cement	4 Frozen fish and seafood	4 Chemicals
5 Paper	5 Steel	5 Others	5 Plastics
6 Cement	6 Aircraft		6 Shoes
7 Fertiliser	7 Motor vehicles and components		7 Soybeans
8 Light engineering products	8 Other machinery and equipment		8 Iron ore

A Manufacturing industry in Bangladesh and Brazil.

Manufacturing industry in Bangladesh grew at a slow but steady rate during the 1990s. Manufacturing increased from 12.9% of GDP in 1990 to 15.4% in 2000. However, the current share of GDP appears to be too small for manufacturing to spearhead sustained high growth of the economy. The growth of Bangladesh's manufacturing sector has also been narrowly based, with readymade garments accounting for nearly a quarter of the growth. Other important export industries contributing to the growth of manufacturing are fish and seafood, and leather tanning. Major import-substitution industries, which grew during this period, include pharmaceuticals, cigarettes, printing and iron and steel re-rolling mills.

3 a) What is meant by *appropriate technology*? (1)

b) Describe two examples of appropriate technology. (4)

c) What is meant by *import-substitution industries*? Give two examples in Bangladesh. (4)

4 For a named industrial city in a developing country:

a) describe and account for the growth of manufacturing in that city (4)

b) describe the problems that industrialisation has posed and the solutions adopted. (4)

Taiwan – A Newly Industrialised Country

The case study on page 147 of the pupil book is on Malaysia, a newly industrialised country (NIC), which is attempting to follow the path taken by the original four tigers of South-East Asia (South Korea, Taiwan, Hong Kong and Singapore) after 1960. This case study on Taiwan, one of the four tigers, shows a possible way forward for Malaysia.

Taiwan is an island 140 km off the coast of China. To all appearances an independent country with its own president, parliament, currency and armed forces, Taiwan is not generally recognised as a sovereign state; effectively it is a self-governing province of China. Taiwan has few natural resources. There are small amounts of copper, oil and coal, but not enough to meet domestic demands. Taiwan was one of the first NICs and is now one of the wealthiest and strongest economies in South-East Asia.

Economic Growth

Taiwan's economic growth since 1949 has been very great. It is the world's 15th largest economy – an astonishing achievement for a relatively small island, with few natural resources. This is explained by the following factors.

- US aid, especially during the 1950s and 1960s, which was aimed at establishing Taiwan as a successful capitalist alternative to mainland China's communism.

- Cheap wages: in 1965 Taiwanese factory workers received ten times less, on average, than their Japanese counterparts.

- A strong government, which under martial law crushed opposition and prevented trade unions from taking strike action.

- Since the late 1980s, a growth in entrepreneurial activity with many local companies flourishing, especially in micro-electronics.

Taiwan had one of the fastest rates of economic growth in the world between 1960 and 1990, with an annual average growth rate of 9.1% for real gross domestic product (GDP). Taiwan passed swiftly through several stages of economic development in the following ways.

- From 1950, the government reformed the agricultural system, reducing the power of the landlords and improving the food supply. Generous compensation to the landlords encouraged them to invest in industrial development. At first the emphasis was on textiles, bicycles and light electronics for the growing domestic market (import substitution) and for export.

- During the late 1960s and into the 1970s, heavy industries such as steel, shipbuilding and petrochemicals were established.

- Consumer electrical products, notably radios and televisions, became increasingly important during the 1970s.

- The emphasis changed again to high-tech electronics during the 1980s, including PCs, CD players, video recorders and micro-electronics.

- Taiwan's economy is now highly diversified and the role of the service sector within the economy has increased rapidly (see Figure A). Taiwan has now lost the advantage of cheap labour and has to depend upon the continuous improvement of its technology, skills and techniques to continue to flourish. Wages in Taiwan are, for example, five times higher than those in the Philippines.

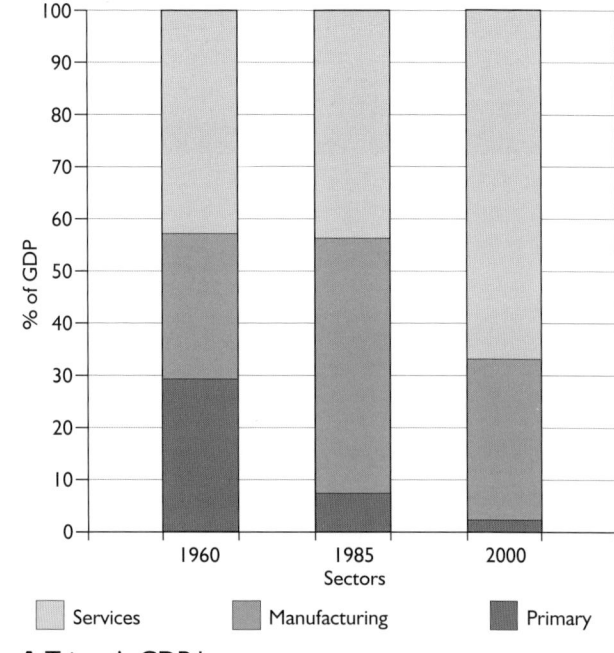

A Taiwan's GDP by sector.

High-Tech Industry

Since the 1980s, Taiwan's IT sector has replaced traditional industries as the main engine for growth and has become the nation's biggest export industry. Taiwan is the world's third largest IT producer, after the USA and Japan. Taiwan's IT manufacturers build almost 40% of the notebook PCs, 67% of the motherboards, keyboards and mice, and almost 60% of the monitors built in the world. Taiwan also supplies over 10% of the world's integrated circuits.

In the year 2000, Taiwan's IT industry produced hardware worth an estimated $33 billion, of which $20 billion was manufactured in the country itself. One Taiwanese company, Acer Computers, is the world's third largest manufacturer of PCs. Acer has two factories in the Philippines and plans five new plants in mainland China. This move to foreign locations has been encouraged by the difficulty of employing production line workers in Taiwan. Wages in Taiwan are at least 15 times higher than on mainland China.

In 1980, the Taiwanese government established a science-based industrial park at Hsinchu, 70 km south-west of Taipei. By 1990, there were 111 high-tech factories on the park, employing 16 000 people and by 2001, this had grown to 440 factories employing 102 745 on a site covering over 600 hectares. Electronics, PCs, telecommunications and biotechnology are the main industrial sectors found on the park (see Figure B on page 108). Twenty per cent of the companies on the science park are foreign-owned. Several transnational companies have plants there, including Philips, IBM and Wang Computers. A number of Taiwanese companies, which have become important players on the world stage, are also located at Hsinchu, including Acer Computers, which has two PC plants on the park. Production from the science park amounts to 10% of Taiwan's gross national product.

The largest single sector on the science park is that of integrated circuits. The companies combine the whole production process including semiconductor materials, integrated circuit design, masks, production, testing and packaging. Taiwan has become one of the most important manufacturers of semiconductors with 7% of total world production by value.

In addition to factory and laboratory buildings, the science park has residential and recreational areas including a swimming pool, and tennis and basketball courts. A lake, trees and flower gardens add to the high quality of life for those living and/or working in the science park. The average age of the workforce in the park is only 31 years old. Returning expatriates have played an important role in the development of the park – 109 of the park's companies were founded by expatriates.

Hsinchu Science Park has had over £425 million worth of government investment in software and hardware facilities. Within the park there are several research facilities employing 12% of the workforce and two universities are located nearby. Three miles east of the science park is the Industrial Technology Research Institute which includes 12 research facilities employing over 6 000 engineers researching for aerospace, chemicals, computers, electronics, communications technologies, opto-electronics (fibre optics, liquid crystal displays (LCDs), light emitting diodes (LEDs)) and environmental protection.

A second science park covering 638 hectares was opened at Tainan in 1998. By 2001, the park already had 37 factories on site.

Industry	Number of firms	Employees
Integrated circuits	173	61 135
Computers and computer peripherals	66	16 064
Telecommunications	71	7 334
Opto-electronics	76	16 225
Precision machinery and materials	19	1 351
Biotechnology	35	636
Totals	440	102 745

B Industries on the Hsinchu science-based industrial park.

Exam Practice Questions

1 Study Figure C.

a) What was the population of Taiwan in (i) 1965 and (ii) 2001? (2)

b) Describe the changes in birth and death rates shown on the graph. (4)

c) Calculate the natural increase in Taiwan's population in (i)1965 and (ii) 2001. (4)

d) How does Taiwan's status as a NIC help to explain the pattern of population change revealed by the graph? (3)

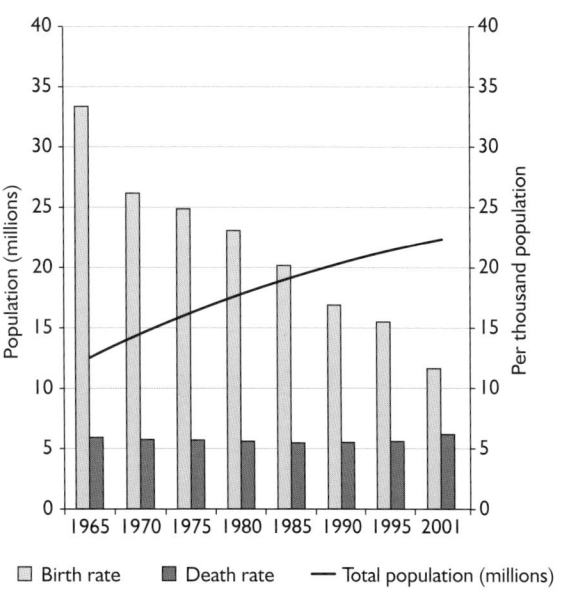

☐ Birth rate ■ Death rate — Total population (millions)

C Population of Taiwan.

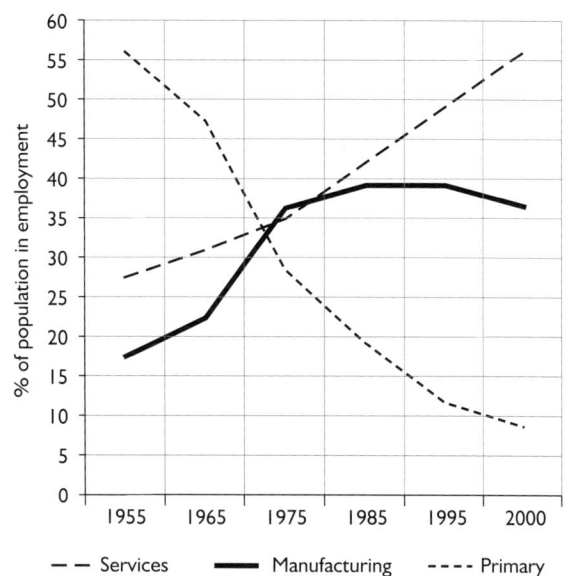

−− Services ▬▬ Manufacturing ---- Primary

D Employment by industrial sector.

2 Taiwan has few natural resources and is one of the world's most densely populated countries. Explain why it has been able to develop such a prosperous economy from such a limited resource base. (4)

3 a) What stages of development did Taiwan's economy pass through between 1950 and 2000? (5)

b) How are these stages shown in the graph of employment by industrial sector in Figure D and in the bar chart of contribution to GDP by sector in Figure A on page 106? (4)

4 Describe the Hsinchu Science Park and account for its success. (6)

Taiwan – A Newly Industrialised Country

These model answers are designed to show you how to get full marks for the longer questions in the Case Study Extra for this section. Read the question carefully and write your own answer. Then read the model answer and the examiner's notes to see what he or she is looking for when awarding the marks. Decide how many marks you think the examiner would give your answer. Decide how to change your answer to this particular question to increase your marks. What will you do differently next time you answer a similar question?

1 d) How does Taiwan's status as a NIC help to explain the pattern of population change revealed by the graph? (3)

> In 1965 Taiwan was at an early stage of its economic development, but during the late 1960s heavy industries were established, followed by consumer electrical products. As better paid jobs in industry became available and as the urban population grew, birth rates fell rapidly. By 2001, Taiwan's birth rate was at the levels expected of a developed country and the increase in the death rate between 1995 and 2001 also indicates a more mature, developed economy.

> **Level 1 (1–2 marks)**: *Includes mention that industrial development led to rapid population increase.*
> **Level 2 (3–4 marks)**: *Demonstrates a clear understanding of the demographic transition and that declining birth rates and low death rates are the result of a wealthier population within a more developed economy.*

2 Taiwan has few natural resources and is one of the world's most densely populated countries. Explain why it has been able to develop such a prosperous economy from such a limited resource base. (4)

> Taiwan was helped by aid from the USA during the 1950s and 1960s, aimed at creating a successful Western-style alternative to communist China. Taiwan had a pool of cheap, well-educated labour, which attracted transnational corporations, and a strong government, which imposed its will on the people, preventing trade unions from taking strike action. Since the late 1980s, there has been a growth in local entrepreneurs establishing industries, especially in high-tech industries such as computers and telecommunications.

> **Level 1 (1–2 marks)**: *Basic answer including reference to high-tech industries not requiring many natural resources.*
> **Level 2 (3–4 marks)**: *Includes important role of aid from the USA in 1950s and 1960s, plus Taiwan's vital human resources, strong government preventing opposition to industrialisation, plus, in more recent times, increased local investment.*

3 a) What stages of development did Taiwan's economy pass through between 1950 and 2000? *(5)*

Stage 1: From 1950, import substitution industries such as textiles, bicycles and light electronics developed as the domestic market grew.

Stage 2: Heavy industries such as steel and petrochemicals followed in the late 1960s under a deliberate government policy to boost the industrial sector.

Stage 3: During the 1970s, consumer electrical products such as radios and TVs became increasingly important, both for export and for the country's growing domestic market.

Stage 4: As Taiwan's industrial expertise grew, so high-tech electronics were established, including computers, CD players and semiconductors.

Stage 5: In recent years, Taiwan's economy has become very diverse with the service sector increasing in importance.

Improve Your Mark!

Level 1 (1–2 marks): *Basic answer, including reference to heavy industries followed by electronics.*
Level 2 (3–4 marks): *Includes more detail of the stages.*
Level 3 (5–6 marks): *Detailed answer, including the shift from import substitution industries, through heavy industries to consumer electronics and now high-tech industries, demonstrating clear understanding of the factors behind each stage.*

b) How are these stages shown in the graphs of employment by industrial sector in Figure D and in the bar chart of contribution to GDP by sector in Figure A? *(4)*

Primary industry accounted for about one half of Taiwan's employment in the 1950s and early 1960s, and almost 30% of GDP. It declined rapidly until by 2000, it accounted for only 8% of employment and 2% of GDP. This reflected the growth in importance of first manufacturing and later services.

Manufacturing industry grew in importance between 1955 and 1985, when it reached its peak share of employment (39%) and contribution to GDP (50%). After 1985, the share of employment remained stable in 1995, but fell to 36% by 2000; the contribution of manufacturing to GDP fell sharply to 31% by 2000. This reflected the increasing importance of services. Employment in services showed steady growth from 27% of the total in 1955 to 35% by 1975, but then increased more rapidly to 56% by 2000, a sure sign of an increasingly developed economy. Services' share of total GDP rose only slightly from 43% in 1960 to 44% in 1985 (the peak of manufacturing's contribution), but then rose dramatically to 67% by 2000.

Improve Your Mark!

Level 1 (1–2 marks): *Basic answer, including decline of primary industry and growth of manufacturing, services and high-tech industries.*
Level 2 (3–4 marks): *Includes more detail, with percentages quoted correctly, and reasons for the growth and recent decline in importance of manufacturing.*

4 Describe the Hsinchu Science Park and account for its success. (6)

The Hsinchu Science Park was established in 1980. It was soon highly successful and hundreds of high-tech companies located there. By 2001, there were 312 factories employing just over 100 000 people. The main industries on the park are micro-electronics, computers, telecommunications and biotechnology. Companies were attracted by the park's extensive and attractive site, by tax concessions and by the skilled and relatively low-paid workforce. They were also attracted by the two universities and the research institute, which are located close to the park. Several transnational companies have chosen Hsinchu, including Philips and IBM. Twenty per cent of the companies on the science park are foreign-owned.

Improve Your Mark!

Level 1 (1–2 marks): *Basic answer, including reference to high-tech industries.*
Level 2 (3–4 marks): *Includes more detail of the type of industries found on the park.*
Level 3 (5–6 marks): *Detailed answer, including reasons for companies choosing to locate to the park, demonstrating a clear understanding.*

10 TOURISM

Chapter 10 is divided into seven sections on tourism. An additional section investigates the Lake District National Park as a case study of the attractions of national parks, the issues created by tourists and some possible solutions. Other case studies in the chapter include the Costa del Sol in Spain, Courmayeur in Italy, a beach village in Barbados and safaris in Kenya.

Recent Trends and Changing Patterns 📖 160 – 161

To introduce this topic, you could use Figures 10.3 and 10.5 on page 161 to count the number of holiday destinations visited by pupils in recent years. Be sensitive to those who may not have had holidays – perhaps the count could also include the holiday destinations of friends or relatives. The countries can be plotted on copies of the world map on page 11.

Activity Sheet 10.1 involves interpretation of bar graphs and manipulation of statistics, using statistics for UK tourism.

Activity Sheet 10.2 deals with the overseas destinations visited by tourists from the UK in 1970 and in 2000. A useful additional activity would be for pupils to conduct their own survey of the whole school to discover the percentage of pupils who have holidayed overseas within the past two years and the destinations.

National Parks in the UK 📖 162 – 163

This section highlights the features of the national parks of England and Wales. It is important that pupils understand that the parks are not owned by the nation – 81% of the land is in private hands. The concept of the 'honeypot' is also an important one. Ask pupils to name any honeypots in their local area – in urban areas these could include any tourist attractions and leisure complexes.

Activity Sheet 10.3 focuses on the location of national parks in the UK, land ownership within the parks and the origins of visitors to the parks.

Coastal Resorts 📖 164 – 165

This section focuses on Spain's Costa del Sol, an area transformed by tourism. It is useful to compare the 19th century tourist boom in Britain, which created towns such as Bournemouth, Scarborough, Southend, Weymouth and Brighton out of fishing villages. Some pupils may have visited the Costa del Sol. Use their experiences to help inform the class of the reality. Websites such as www.costaguide.com provide further detailed information.

Activity Sheet 10.4 covers coastal resorts and contrasts the monthly distribution of foreign visitors to the UK and Spain.

Mountain Resorts 📖 166 – 167

This section studies the Italian alpine resort of Courmayeur. Again, the attractions of the resort are detailed and the benefits and disadvantages that tourism has brought are listed. The attractions can be seen in websites such as www.courmayeur.com; www.news.bbc.co.uk/1/hi/sci/tech/1515276.stm provides further details on the problems. The Alps attracted 120 million visitors in 2001, making them one of the world's most tourism-intensive regions.

Activity Sheet 10.5 includes some useful statistics on tourist arrivals in Switzerland, by country of origin. Germany, the USA and the UK account for half of the total number of tourists. Pupils are required to draw a flow-line map for which they will need copies of the map of Europe on page 9 of the *Teacher's Resource Book*.

Tourism in Developing Countries 📖 168 – 169

Following an introduction including ecotourism, this section studies a West Indies beach village resort. The long lists in Figure 10.27 on page 169 could be subdivided by pupils into three groups: environmental, cultural and economic. This is a good point at which to read the poem 'When the Tourists Flew In' on page 115 of the *Teachers Resource Book*, since it was inspired by a trip to the West Indies and deals with the effects of mass tourism.

Tourism and the Environment 📖 170 – 171

Kenya provides an excellent example of a developing country exploiting its tourist potential to the full. Tourism is Kenya's major source of overseas income. One million tourists visited Kenya in 2000, compared with 350 000 in 1980. The vulnerability of an economy based on tourism is shown by the decline in numbers resulting from terrorist attacks in Nairobi in 1998 and Mombasa in 2002.

Case Study 10 The Lake District National Park 📖 172 – 175

This detailed study of the Lake District highlights the area's attractions and the problems created by the remarkable annual total of 12 million visitors. The 1991 Census data can be updated by visiting the Lake District National Park Authority's website at www.lake-district.gov.uk. Another interesting website is 'Uldale on the Internet' at www.whitestreet.freeserve.co.uk/uldale/stats

Case Study Extra Thailand

As a long-haul destination, Thailand makes a fascinating study. Ecotourism has gained ground in the country as some of the disadvantages of mass tourism have become apparent.

Figure A shows the number of holidays taken by UK residents between 1971 and 2001. In contrast to Figure 10.4 on page 161 of the pupil book, these graphs show holidays of four nights or more.

1 a) What percentage of UK residents took two or more holidays in (i) 1971, (ii) 1981 and (iii) 2001? *(3)*

   b) How has the pattern of numbers of holidays taken changed between 1971 and 2001? *(4)*

   c) What factors help to explain this growth? *(6)*

2 Why has the percentage of UK residents taking no holidays remained almost unchanged? *(2)*

3 What types of holidays are excluded from the bar graph but included in Figure 10.3 in the pupil book? How would their inclusion change the pattern of the bar graphs? *(3)*

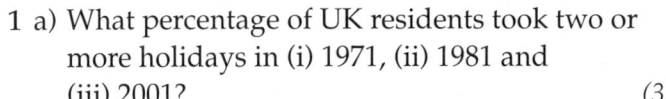

A Holidays taken by UK residents 1971, 1981, 1991 and 2001.

Legend: ■ 2 or more holidays ▨ 1 holiday only ▢ No holidays

4 The table below shows the UK regions visited by UK and overseas tourists in 2001.

   a) Complete the final column showing the difference between UK and overseas tourists. *(3)*

   b) (i) Name the three regions showing the greatest differences. *(3)*

      (ii) Why do you think these regions attract different percentages of UK and overseas tourists? *(6)*

      (iii) Why did so few tourists visit Northern Ireland? *(2)*

Regions	UK residents (%)	Overseas residents (%)	Difference
South-west	25	9	17
Scotland	11	10	1
Wales	9	6	3
Southern	11	12	1
East England	10	9	3
Yorkshire and Humberside	9	5	4
North-west	5	8	3
South-east	7	14	
Cumbria	3	1	
Midlands	5	13	
Greater London	2	9	
North-east	2	3	
Northern Ireland	1	1	

1 Study Figure A.

    a) How many of the attractions appear in the top twelve in both years? *(1)*

    b) Name two attractions that showed an increase in visitor numbers. *(2)*

    c) Name three attractions that dropped out of the top twelve in 2001. *(3)*

    d) Name three attractions that entered the top twelve in 2001. *(3)*

    e) Describe the changing trends in the main tourist attractions in the UK between 1983 and 2001. *(6)*

1983	Attraction	Number of visits (millions)
1	Blackpool Pleasure Beach	6.6
2	The Science Museum, London	3.3
3	The British Museum, London	3.3
4	The National Gallery, London	2.9
5	Westminster Abbey, London	2.8
6	The Natural History Museum, London	2.5
7	Tower of London	2.2
8	St Paul's Cathedral, London	2.0
9	Canterbury Cathedral, Kent	2.0
10	York Minster	2.0
11	Madame Tussaud's Waxworks, London	2.0
12	The Victoria & Albert Museum, London	1.8
		Total 33.4

2001	Attraction	Number of visits (millions)
1	Blackpool Pleasure Beach	6.5
2	The National Gallery, London	4.9
3	The British Museum, London	4.8
4	The London Eye	3.9
5	The Tate Modern Gallery, London	3.6
6	Alton Towers Theme Park	2.5 (est.)
7	The Tower of London	2.1
8	Pleasureland Theme Park, Southport	2.1
9	The Eden Project, Cornwall	1.8
10	The Natural History Museum, London	1.7
11	Legoland, Windsor	1.7
12	York Minster	1.6
		Total 37.2

A The twelve most visited tourist attractions in the UK in 1983 and 2001.

2 Study Figure B.

a) The two most popular destinations have not changed. Name them. *(2)*

b) Why have they remained popular with tourists from the UK? *(4)*

c) Which country shows the greatest increase between 1970 and 2000? *(1)*

d) What factors explain the growth of tourism to that country from the UK? *(5)*

e) How does the pattern of tourist destinations differ between 1970 and 2000? *(3)*

f) Why has the pattern changed? *(5)*

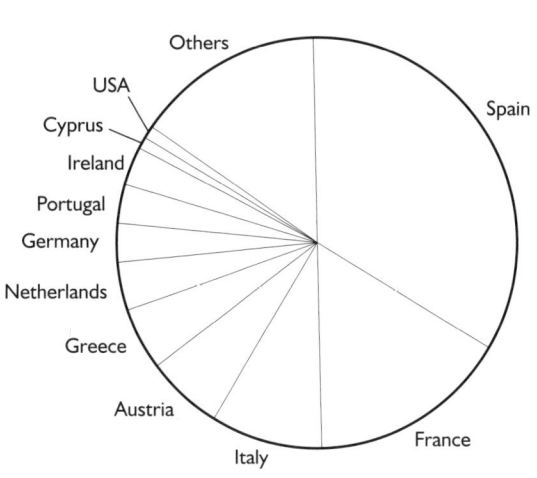
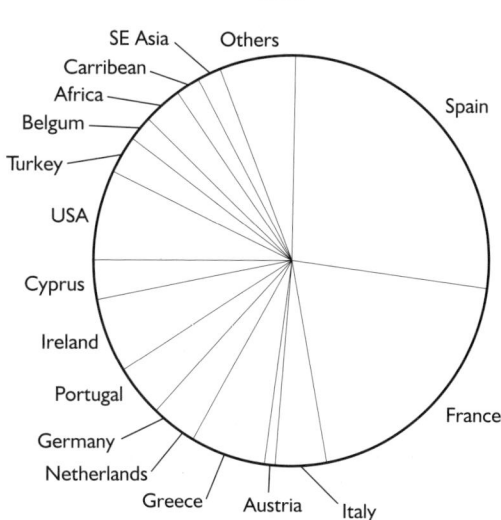

B The overseas destinations visited by tourists from the UK in 1970 and 2000.

1 a) Why were national parks established in the UK? *(3)*

  b) When and where was the first national park in Scotland created? *(2)*

2 Study Figure A.

  a) Use both maps to describe the distribution of national parks in England and Wales. *(4)*

  b) Which national park is surrounded by the most conurbations? *(1)*

  c) Which two conurbations have the shortest motorway link to the Lake District? *(2)*

3 Figure B shows the land ownership of the national parks in England and Wales. How does the land ownership of the Lake District National Park differ from the overall pattern? (See Figure 10.36 on page 173 of the pupil book.) Why doyou think this difference occurs? *(5)*

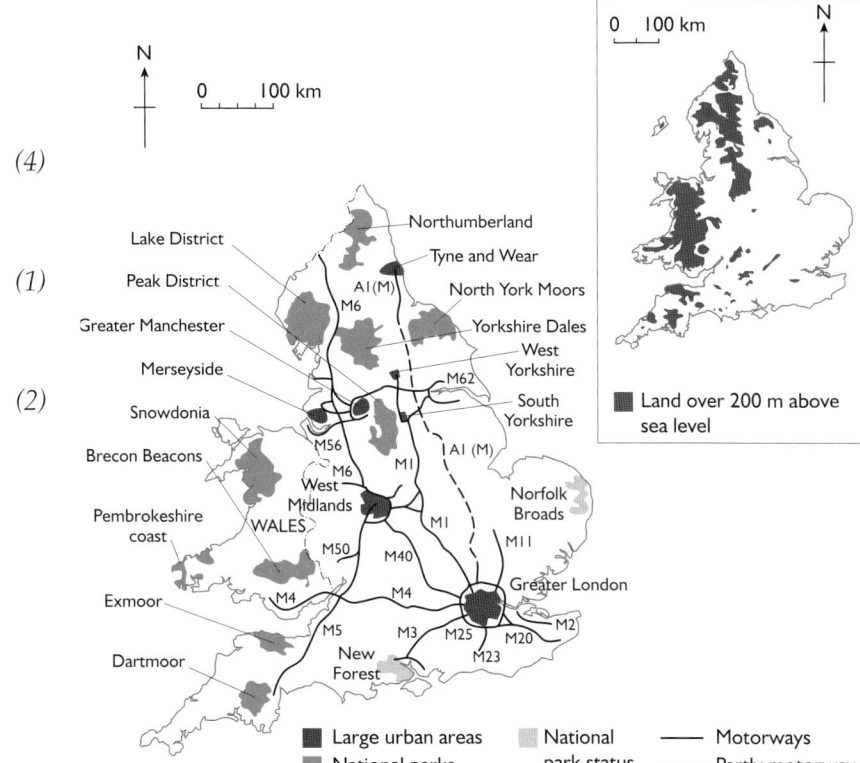

A (i) National parks and (ii) relief over 200 m in the UK.

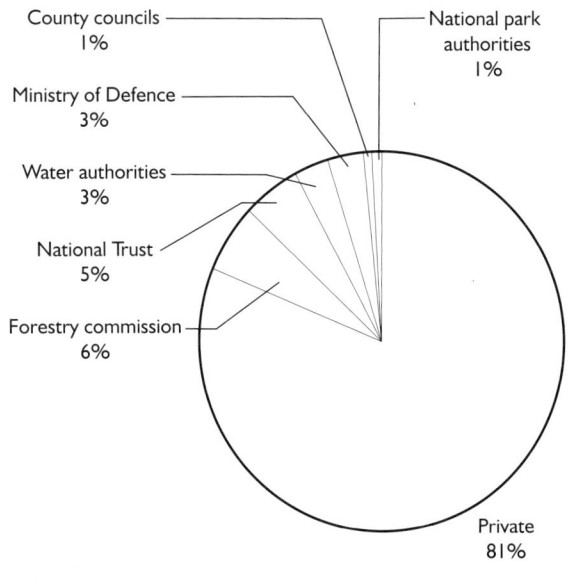

B Land ownership in the national parks in England and Wales.

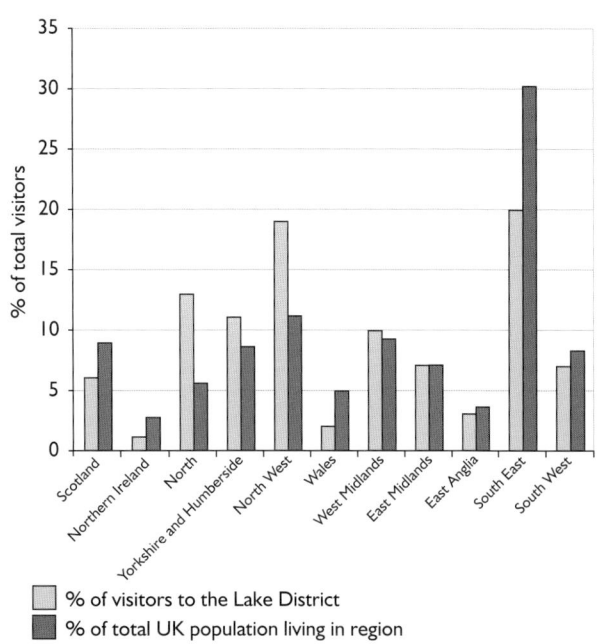

C The origins of holiday visitors to the Lake District.

4 Figure C shows the origin of holiday visitors to the Lake District from each UK region and the percentage share of the total UK population living in each region. Describe and explain the differences in the two sets of figures for (i) the North West, (ii) the South East and (iii) the West Midlands. *(6)*

1 What are the main attractions of a holiday in a coastal resort? (3)

2 Study Figure A.

a) Draw a line graph to show the average hours of sunshine per day for London and Malaga. (6)

b) How does your graph help to explain the attraction of the Costa del Sol for British tourists? (2)

c) An increasing number of Britons visit the Costa del Sol for winter breaks.

(i) What sort of weather can they expect? (2)

(ii) How is this more attractive than the weather in London? (2)

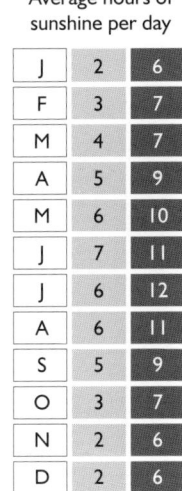

Average hours of sunshine per day

	London	Malaga
J	2	6
F	3	7
M	4	7
A	5	9
M	6	10
J	7	11
J	6	12
A	6	11
S	5	9
O	3	7
N	2	6
D	2	6

☐ London
■ Malaga

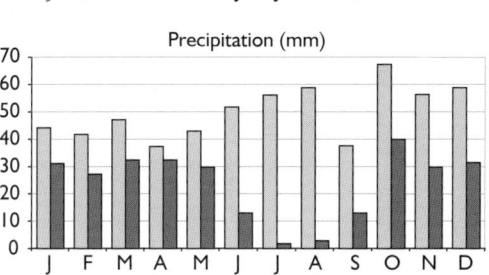

Average daily maximum temperature (°C)

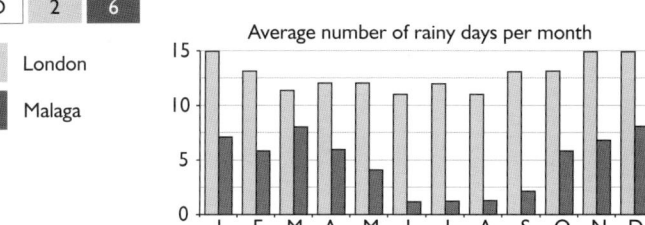

Precipitation (mm)

Average number of rainy days per month

3 a) Study Figure B. Complete the graph by plotting the figures for August to December for Malaga. (5)

b) Describe and try to explain the different patterns shown by the graphs. (6)

4 What has been the environmental impact of mass tourism on the Costa del Sol? (5)

A Climate graphs for the Costa del Sol and London.

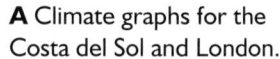

% of foreign visitors

- - - Spain ▬▬▬ United Kingdom

Months	UK	Spain
J	6	4
F	6	3
M	7	5
A	9	9
M	9	7
J	9	10
J	11	15
A	12	21
S	9	13
O	8	5
N	7	3
D	7	5

B Monthly distribution of foreign tourists visiting the UK and Spain.

1 What factors make the Alps popular for winter holidays? (4)

2 For one named Alpine resort, state what (i) benefits and (ii) disadvantages the increase in tourism has had on the resort. (6)

3 The development of tourism in mountainous areas can have a major effect on the environment and local economy. How can planners ensure that mountain areas are sustainably developed? Use an example of a mountainous area you have studied. (5)

4 Study Figure A.

a) On an outline map of Europe, draw flow lines to show where the tourists visiting Switzerland from other European countries come from. Use the scale of 1 mm to represent 1%. (6)

b) Why do you think that few tourists come from Switzerland's neighbours, Austria and Italy? (1)

Country of origin	% of total foreign tourists	Country of origin	% of total foreign tourists
Germany	30	Belgium	3
USA	11	Spain	2
UK	9	Austria	2
Japan	7	Israel	1
France	7	Sweden	1
Italy	6	Others	17
Netherlands	4		

A Tourist arrivals in Switzerland in 2001 by country of origin.

5 Study the article on slow change in Switzerland.

a) Overnight stays in Switzerland were lower in 2001 than they had been in 1990. What factors help to explain the decline? (4)

b) What could Swiss planners do to boost tourism in the Swiss Alps? (4)

Slow change in Switzerland

Switzerland has always been a magnet for tourists. Its mixture of grand Alpine peaks and peaceful lakes, comfortable family-run hostelries and efficient mountain railways has made it one of Europe's most popular tourist destinations. However, the rapid growth in Switzerland's tourist industry, which marked the 1950s and 1960s, came to a halt a long time ago.

For most of the 1990s, Switzerland's tourist industry declined. However, the six-year fall in overnight stays was stemmed in 1997 when stays rose by 3.7%, to 30.8 million. By 2001, they had risen another 10% to 33.9 million. But the figure is still 8% below its 1990 peak of 35.6 million and also below the level of 30 years ago.

Switzerland is far more reliant on tourism than most European countries – it accounts for 5.6% of the country's gross domestic product. Switzerland has suffered a big loss of market share since the 1970s. Part of this is unavoidable as tourist preferences have changed in traditionally important markets such as Germany and France, but part of it also reflects the slow pace of change and under-investment in Switzerland's tourist industry. There is too much old hotel capacity, which is preventing the building of more modern facilities which appeal to younger visitors. Switzerland's old hotels are part of its undeniable charm, but they are catering to an increasingly older clientele.

Tourism in Thailand

Thailand is one of the long-haul destinations that have become increasingly popular with British tourists who seek something more exotic than the Mediterranean can offer. Thailand's tropical rain forests, spectacular beaches and unique culture are an attractive mixture. Thailand makes an interesting contrast with the case studies in the pupil book; moreover, eco-tourism (or green tourism) has become increasingly important there.

The Importance of Tourism

Tourism is very important for Thailand. It has played a major role in Thailand's economic development for over 40 years. Since 1990, Thailand has been placed among the top 20 most popular tourist destinations in the world (see Figure A). Tourist numbers increased from 5.3 million in 1990 to 10.1 million in 2001, a more rapid growth than average for the top 20 destinations. In 2001, tourism earned Thailand £4.5 billion and accounted for 5.4% of the country's GDP.

Rank in 2000	Country	1990	2000	Difference	% difference
1	France	52.5	75.5	23.0	44
2	China (incl. Hong Kong)	19.5	51.0	31.5	162
3	USA	39.4	50.9	11.5	29
4	Spain	34.1	48.2	14.1	41
5	Italy	26.7	41.2	14.5	54
6	UK	18.0	25.2	7.2	40
7	Russia	7.2	21.2	14.0	194
8	Mexico	17.2	20.6	3.4	20
9	Canada	15.2	20.4	5.2	34
10	Germany	17.0	19.0	2.0	12
11	Austria	19.0	18.0	-1.0	-5
12	Poland	3.4	17.4	14.0	412
13	Hungary	20.5	15.6	-4.9	-24
14	Greece	8.9	12.5	3.6	40
15	Portugal	8.0	12.0	4.0	50
16	Switzerland	13.2	11.4	-1.8	-14
17	Netherlands	5.8	10.2	4.4	76
18	Malaysia	7.5	10.2	2.7	36
19	Turkey	4.8	9.6	4.8	100
20	Thailand	5.3	9.5	4.2	79
Totals		343.2	499.6	156.4	46

A International tourist arrivals (in millions) in 1990 and 2000.

Bangkok is the main tourist centre. Its spectacular palaces and temples are very popular and its extensive shopping and nightlife are also highly attractive to tourists. Outside Bangkok, the main tourist centres are the beach resorts of Pattaya and Phuket, and the northern inland resort of Chiang Mai, the ancient capital of Siam (as Thailand was known before 1945).

Sixty-five per cent of Thai people are self-sufficient rice farmers. The Thai government is keen to develop tourism in order to earn foreign exchange and create wider job opportunities. In 2001, the Tourism Authority of Thailand stated:

'To diversify tourist attractions, lesser known destinations and new attractions must be developed and introduced to visitors as new options for them. In this connection, each community will be urged to bring out what it has. The project to diversify tourist attractions has already started in the north-east of Thailand. The Tourism Authority of Thailand is stepping up the public relations campaign for tourism promotion in the north-eastern region, which is in fact ready to host more visitors in terms of accommodation, convenient travel and tourist sites. More than 50 new tourist destinations in the north-east have been included in the tour programmes of various tour agencies. In 2001, Thailand recorded about 10.1 million tourist arrivals. Of that number, only 3% went to the north-east. With strong promotion, it is expected that the number of international tourists visiting north-eastern Thailand will double within three years.

In the dim economic situation, tourism is regarded as a bright spot to earn additional income for Thailand. The development of new tourism destinations will ensure the growth of the industry and spur the country's economic growth.'

Ecotourism

This desire to spread the influence of tourism across Thailand is linked with concerns over the negative impact which tourism can have. This has led to the growth of ecotourism. The definition of ecotourism used by the International Ecotourism Society is 'responsible travel to natural areas that conserves the environment and improves the welfare of local people'. In her book *Ecotourism and Sustainable Development*, the author Martha Honey gives seven points to define ecotourism:

- involves travel to often remote natural destinations
- minimises impact
- builds environmental awareness
- provides direct financial benefits for conservation
- provides financial benefits and empowerment for local people
- respects local culture
- supports human rights.

The extract on page 121, entitled 'Making a difference', gives an account of an ecotourist holiday in Thailand. Such holidays are increasing in importance and new resorts are being constructed in order to cater for them. One example is at Chumporn Cabana, a new development in southern Thailand, 500 km south of Bangkok. The company building the resort professes ecotourist ideals: 'we hope to take the tourism industry in a different direction, one which will preserve and enrich our beaches and forests without sacrificing the wellbeing and integrity of the communities which have made them a favoured destination for Thais and tourists alike.' The resort features:

- an environmentally friendly water treatment system
- an organic waste treatment facility which will produce compost to be given to local farmers
- local, sustainable materials which were used in constructing the buildings; locally-made bricks were used instead of rare hardwood timber
- a low-energy form of temperature control using water cooled walls rather than energy guzzling air conditioning
- no development on the beach itself, including no beach furniture or water sports facilities
- direct purchase of food from local farmers, provided they use organic methods of production.

The trend towards ecotourism holidays presented as sustainable, nature-based and environmentally friendly is now the subject of considerable controversy. Ecotourism is the tourism industry's fastest growing subsector, with an estimated worldwide annual growth of 10–15%. Governments, as well as the tourism industry, promote ecotourism with its claims of economic and social sensitivity. However, there are concerns that this is often simply a marketing tactic. There has been growing opposition from Thai environmentalists and villagers to the move of the government to open up protected areas for mass ecotourism (see the extract from the responsibletravel.com website below). Ecotourism may actually have a more ruinous effect than mass tourism because it moves into highly sensitive environments such as nature reserves, virgin forests, wetlands and other wilderness areas. By contrast, mass tourism concentrates its impact on already developed resorts.

Aware of the threat which tourism poses to the Thai culture and environment, the Tourism Authority of Thailand launched the Green Leaf Programme in early 1997. It gives due recognition to hotels which meet certain international criteria for environmental preservation. The hotels are audited on the basis of environmental standards, waste management, energy and water efficiency, purchasing policy, air and noise pollution, water quality, storage and management of fuel, gas and toxic waste, ecosystem impact and community cooperation. The programme is also supported by the United Nations Environment Programme (UNEP), which considers this to be a pilot project for the Asian and Pacific region. The programme has been linked since 2002 with the Australian-based Green Globe 21 programme, which provides a benchmarking service in support of ecotourism.

It is important that tourists respect the culture and traditions of the societies they are visiting. Some Western tourists have not done this. Topless sunbathing, excessive drinking, violent and indecent behaviour have given offence and a poor impression of Westerners to people in developing countries such as Thailand. This may have had some influence on the actions of international terrorists. Terrorist organisations opposed to the spread of Western culture and influence have deliberately attacked tourists. Nearly two hundred were killed in a bomb attack in Indonesia in 2002; earlier murderous attacks on tourists took place in the Philippines and Egypt. If ecotourism succeeds in fostering a more meaningful relationship between the tourists and residents with less social and cultural disruption, it may have a role to play in improved understanding between peoples.

Making a difference

Explore the vibrant cities of Bangkok and Chiang Mai and the infamous Golden Triangle; visit some of Thailand's most interesting and beautiful ancient sights; spend the night in a Thai village as guests of a local family; learn to cook the mouth-watering dishes that have made Thai cuisine a world favourite. On this exciting trip you do all of this and more on a fabulous journey to the heart of northern Thailand.

This trip gives travellers a wonderful opportunity to experience life outside of Thailand's bustling modern cities. We stay in small local hotels, employ many local guides and representatives and remind travellers of the need to be environmentally responsible. Travellers are also encouraged to shop for local handicrafts in the vibrant night markets and maybe indulge in a traditional Thai massage.

Just north of Chiang Mai we spend a night as guests of a local family in their traditional teak house where the money from our visits goes directly to the family. In addition to the inevitable interaction between hosts and travellers, we also become 'students' for the day as the family give us a real 'hands on' cooking lesson – starting with shopping in the local market for fresh ingredients, and picking herbs and spices from the family garden. The following day there is an opportunity to visit the local temple to gain an insight into Buddhism and the role of the temple in village life. We also visit the Elephant Conservation Centre at Lampang, where we learn about the Centre's efforts to bring them back into the wild and have the opportunity to contribute financially to the preservation of these beautiful animals by taking an elephant back ride.

Exam Practice Questions

1 a) What are the attractions of Thailand for international tourists? *(4)*

b) Describe two advantages of tourism for Thailand. *(3)*

2 a) What do you understand by the term *ecotourism*? *(2)*

b) Using examples, explain how ecotourism benefits the environment and local people. *(6)*

3 Study the extract entitled 'Opposition to ecotourism', taken from an essay by the Thai activist, Ing K, and the poem 'When the tourists flew in'. Local people may have different opinions about the growth of tourism in their area. Suggest two reasons why local people may (i) welcome or (ii) oppose tourist developments. *(4)*

Opposition to ecotourism

Land speculation became a national pastime, permeating every beautiful village, however remote. Land prices skyrocketed. Villagers sold agriculturally productive land to speculators. Practically overnight, fertile land became construction sites. National parks and forest reserves were encroached upon by golf courses and resorts. Greed and consumerism devastated whole communities all over Thailand. Diverse local, social and economic activities were replaced by an ecotourism monoculture.

Contrary to claims, local people do not necessarily benefit from ecotourism. Tourism-related employment is greatly overrated: locals are usually left with low-paid service jobs such as tour guides, porters, and food and souvenir vendors. In addition, they are not assured of year-round employment: workers may be laid off during the off-season.

Most money, as with conventional tourism, is made by foreign airlines, tourism operators and developers who send the profits back to their own economically more developed countries. In the end, we have nothing to show for it but whole graveyards of unsold high-rise apartments, golf course and resort developments and housing estates.

When the tourists flew in *by Cecil Rajendra*

When the tourists flew in
The Finance Minister said
'It will boost the Economy
The dollars will flow in.'

The Minister for the Interior said
'It will provide full
and varied employment
for all the indigenes.'

The Minister of Culture said
'It will enrich our life …
contact with other cultures
must surely
improve the texture of living.'

The man from the Hilton said
'We will make you a second
Paradise;
for you it is the dawn
of a glorious new beginning!'

When the tourists flew in
our island people
metamorphosed into
a grotesque carnival
a two week side-show.

When the tourists flew in
our men put aside
their fishing nets
to become waiters
our women became whores

When the tourists flew in
what culture we had
flew out of the window
we traded our customs
for sunglasses and pop
we turned sacred ceremonies
into ten-cent peep shows.

When the tourists flew in
local food became scarce
prices went up
but our wages stayed low

When the tourists flew in
the hunger and the squalor
were preserved
as a passing pageant
for clicking cameras
a chic eye-sore!

When the tourists flew in
we were asked
to be 'side-walk ambassadors'
to stay smiling and polite
to always guide
the 'lost' visitor …
Hell, if we could only tell them
where we really want them to
go!

Cecil Rajendra is a Malaysian poet and lawyer whose poetry about peace, justice and ecology has been translated into many languages. This poem was written after visits to Trinidad and Haiti, but it applies to tourism in any developing country.

4 Describe and explain Thailand's Green Leaf programme. *(4)*

Tourism in Thailand

These model answers are designed to show you how to get full marks for the longer questions in the Case Study Extra for this section. Read the question carefully and write your own answer. Then read the model answer and the examiner's notes to see what he or she is looking for when awarding the marks. Decide how many marks you think the examiner would give your answer. Decide how to change your answer to this particular question to increase your marks. What will you do differently next time you answer a similar question?

2 b) Using examples, explain how ecotourism benefits the environment and local people. (6)

> Ecotourism is environmentally friendly since it aims to conserve the environment and increase environmental awareness. Income can be used to fund improved environmental protection. Food can be directly purchased from local farmers and local residents can be employed to support and service the tourists. Ecotourists are thought to spend considerably more than mass tourists and for developing countries, having people visit, look at things that require minimal investment and pay lots of money for the privilege, can seem to bring nothing but benefit. Chumporn Cabana is a resort in southern Thailand. The buildings were constructed out of local brick rather than rare hardwood timber. The resort includes an environmentally friendly water treatment system, and an organic waste treatment plant, which produces compost that is given to local farmers. Produce is then purchased from the farmers. No development is allowed on the beach itself.

> **Level 1 (1–2 marks)**: *Basic answer including reference to income for local people.*
> **Level 2 (3–4 marks)**: *Includes beneficial environmental effects.*
> **Level 3 (5–6 marks)**: *Detailed answer, including a named example, demonstrating clear understanding.*

3 Study the extract entitled 'Opposition to ecotourism', taken from an essay by the Thai activist, Ing K, and the poem 'When the tourists flew in'. Local people may have different opinions about the growth of tourism in their area. Suggest two reasons why local people may (i) welcome or (ii) oppose tourist developments. (4)

> Local people may welcome tourist developments because the tourists will spend money purchasing local produce and locally-made souvenirs. Local people will also seek employment generated by tourism, in places such as hotels, shops and restaurants. Some local people may oppose tourist developments because the tourists may behave inappropriately, showing no respect for the local culture. Tourism may also damage the local environment through construction of roads and hotels.

> **Level 1 (1–2 marks)**: *Basic answer including reference to earning money or gaining employment, and, in opposition, damage to the environment.*
> **Level 2 (3–4 marks)**: *More detailed answer including increased income into the local economy and increased employment opportunities; opposition due to environmental damage through construction work and through inappropriate behaviour by the tourists.*

CHAPTER 11 covers the topics within World Development and Interdependence in five sections followed by an extended case study comparing and contrasting Japan and Kenya.

Patterns and Characteristics of Development 180 – 181

The concept of the North-South divide and meaning of the terms MEDCs and LEDCs begin this chapter, followed by a discussion of the contrasting characteristics of these two groups of countries. Indicators of development, both economic and social, are discussed and summarised in Figure 11.4 on page 181.

Activity Sheet 11.1 tests and extends these ideas. It takes the indicators of 'adult literacy rate' and 'per capita calorie intake as % of daily requirements' to consider the level of development of a country. It also introduces the benefits of scattergraphs and Spearman's Rank Correlation Coefficient for studying the correlation of two sets of development data. Pupils are also encouraged to consider other possible indicators of development, using Figure 11.3.

Inequalities in World Development 182 – 183

The Human Development Index (HDI) is explained and its effectiveness as a development indicator shown. A large number of statistics are given in Figure 11.5 on page 182 for individual pupil and class use. Pupils might describe the choropleth map in Figure 11.6 on page 183 as a useful skills exercise before attempting to explain the pattern.

Activity Sheet 11.2 uses Mexico as an example of how improvements can be seen in the HDI. It also links the HDI to the North-South divide.

Sustainable Development and Appropriate Technology 184 – 185

The difference between 'quality of life' and 'standard of living' is explained. Sustainability is considered in both MEDCs and LEDCs, using Ladakh in India to illustrate multiple source sustainable energy production in a particularly difficult environment.

Activity Sheet 11.3 uses the northern Kenyan settlement of Korr to illustrate how appropriate technology can help to reverse environmental damage resulting from overexploitation. Attitudes to the scheme are explored using a role play exercise.

World Trade and Interdependence 186 – 189

This is often not an easy section of the specifiacation for many pupils, as the subject is beyond their experience of the world to date. The section is therefore divided into clear stages on Patterns of world trade, Free trade, tariffs and quotas, Trading groups, Direction of world trade, World trade as a measure of development, GATT (General Agreement on Trade and Tariffs) and the WTO (World Trade Organisation). Tables and pie charts are used to summarise data.

Activity Sheet 11.4 supports this section by utilising the data shown in Figure 11.11 on page 186, but involving the different skill of constructing split bars. Question 4 asks pupils to compare the uses of split bars and pie charts.

Aid 190 – 191

Types of aid are categorised in Figure 11.17 on page 190 and then analysed, along with discussion on the advantages and disadvantages for the recipient country.

Activity Sheet 11.5 extends this discussion to the donor countries as well. It also looks at one type of aid, Official Development Assistance, in the context of Africa, using a choropleth map. This activity sheet is directed at more able pupils.

Case Studies 11a and 11b Japan and Kenya 192 – 195

The theoretical points on both MEDCs and LEDCs made throughout the chapter are illustrated and compared under two headings, Levels of development and Trade and interdependence.

Case Study Extra One Form of Aid – Sponsoring Children in LEDCs

This is concerned with a further branch of aid, this time in the voluntary sector. The topic is taken beyond what would be needed in an exam, but extends the more able pupil. Two alternative model answers are suggested to the same exam practice question to show pupils that it is not the decision they make which gets marks, but the reasons they give to support it.

An indicator of development is a measure or statistic that tells us how developed a country is. Such indicators have been used for many years and were the main sets of statistics used before the Human Development Index (HDI) (see pages 182–183 in the pupil book) was devised. Study Figure A, which gives data for two indicators of development.

Country	% of adult population who are literate, 1998	Per capita calorie intake as % of daily requirements, 1998
UK	99	130
USA	99	138
India	85	101
Guatemala	55	103
Zambia	73	87
Peru	85	87
Jamaica	98	114
Rwanda	50	82
Afghanistan	29	72
Uganda	48	93

A Literacy rates and calorie intake per capita in 1998 by country.

1 a) A scattergraph to show the links between these two sets of data has been started for you in Figure B. Complete the graph using the rest of the data in Figure A. *(7)*

b) Draw a best fit (trend) line. *(2)*

c) What does your best fit line show about the relationship between the two data sets? *(2)*

d) Is this relationship a positive or a negative correlation? *(1)*

e) An anomaly is a point that is noticeably further away from the best fit line than the other points. Name the countries that are anomalies on this graph. Give a reason for these anomalies. *(4)*

f) Explain why, in general, the higher the percentage of the adult population who are literate in a country, the higher the per capita calorie intake. *(4)*

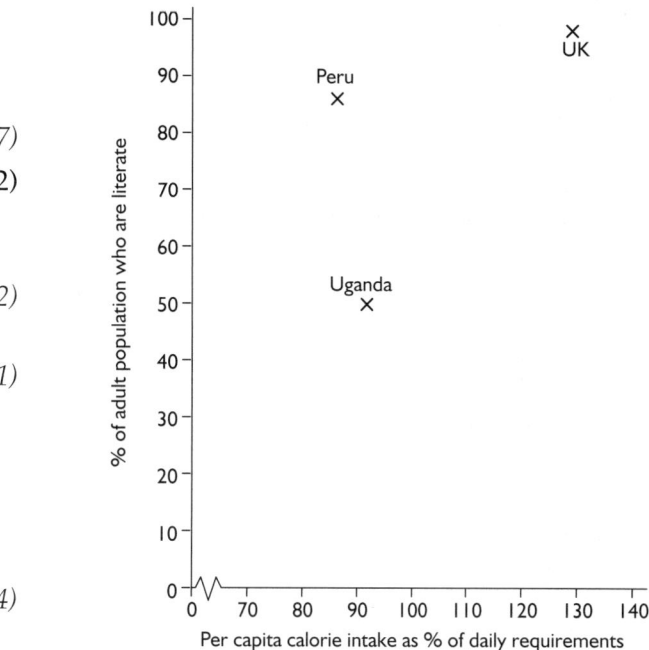

B The relationship between literacy and nutrition.

2 a) Name six other indicators of development. (Figure 11.3 on page 181 of the pupil book gives some suggestions.) Do not include the HDI in your list, because it is made up of several indicators. *(6)*

b) Select any pair of indicators of development. Do you expect to find a relationship between them? Explain why. *(4)*

c) Test your answer by drawing another scattergraph or by using the data sets in a Spearman's Rank Correlation exercise. Comment on the meaning of the results. *(10)*

d) If you used the Spearman's Rank Correlation method, say whether or not you prefer it to the scattergraph method. Give your reasons. (There is no correct answer, it is simply a matter of opinion. The marks are allocated for the quality of your reasons.) *(3)*

1 a) Define what is meant by the term *North-South divide* (see Figure 11.1 on page 180 in the pupil book). *(2)*

   b) Name two countries on each side of the North-South divide. *(4)*

2 a) Define what is meant by the term *Human Development Index*. *(2)*

   b) Name the three indicators of development that make up the Human Development Index (HDI). *(3)*

   c) State whether the HDI is likely to be high or low on each side of the 'divide'. *(2)*

3 Read the article on Mexico and answer the following questions.

   a) Which illness have the Mexicans been fighting in this health programme? *(1)*

   b) Explain how the Mexican government encouraged its people to overcome this illness. *(4)*

   c) How many children would have died without the government programme? *(1)*

Mexico: 30 000 saved since 1990

In three years, Mexico has halved child deaths from diarrhoeal disease. An independent evaluation completed in September 1994 shows a 56% fall in diarrhoea deaths among under-five deaths (1990 to 1993). This means that approximately 30 000 young lives have been saved so far as a result of Mexico's attempt to achieve the 1995 goal of 80% ORT use (see below) set by the 1990 World Summit For Children.

In the four years following the Summit, President Carlos Salinas de Gortari gave particular support to the goal of educating all Mexican families in the use of oral rehydration therapy (ORT), the simple and low-cost technique that can prevent most deaths from diarrhoeal disease.

By the end of 1993, over 5 million mothers in Mexico had been trained in ORT. In the worst affected areas, the Ministry of Health has trained approximately 1 million women as health representatives – able to teach others how to prevent and treat the dehydration which turns ordinary diarrhoea into a killer disease.

National oral rehydration days and child health weeks have helped to spread these basic messages to virtually every village and urban neighbourhood in Mexico. In 1993 alone, ORT 'advertisements' appeared 120 000 times on national television and more than 2.3 million times on radio. In the same year, almost 8 million posters, pamphlets, and leaflets carried the ORT message under the title 'The best solution'.

To meet the increased demand generated by these campaigns, Mexico's annual production of oral rehydration salts has increased from 9 million packets in 1989 to 83 million in 1993.

Source: **The State of the World's Children** (*UNICEF, 1995*).

4 Suggest four ways in which the actions of the Mexican government could improve the country's HDI. *(4)*

5 Using one or more countries you have studied, discuss how people's standard of living or HDI has improved. *(9)*

1 Look up the term *sustainable development* and write down the definition. (2)

2 a) Read the article about Korr and study Figure A(i). The missionaries made some important mistakes, which led to serious environmental degradation. Make a table of the mistakes and their consequences. (5)

 b) Figure A(ii) shows Korr after a sustainable development scheme had been launched by a charity. The labels for the second diagram are listed below. Write the correct label in each empty box. (3)

 A Areas fenced off to allow grass to regrow.
 B Tree planting programme
 C People encouraged to settle at lower population densities to decrease environmental pressure.

The settlement at Korr in northern Kenya

Korr has long had a number of boreholes that have been used for generations by nomadic people and their herds. Water was only taken seasonally, so groundwater levels did not drop from year to year. Pasture was grazed, but not overgrazed.

Then a Roman Catholic mission was set up at Korr. To support this new small permanent population, a deeper borehole was dug, providing water reliably all year round. This improved water supply and other services, e.g. access to health care and education.

Because of the increased availability of water, some nomads decided to give up their traditional way of life and become sedentary (living in one place, instead of travelling). Nevertheless, these people still earned their living through their livestock and the animals placed an extra pressure on this fragile environment. The people also took more resources from the environment than previously, simply because they were there all year round.

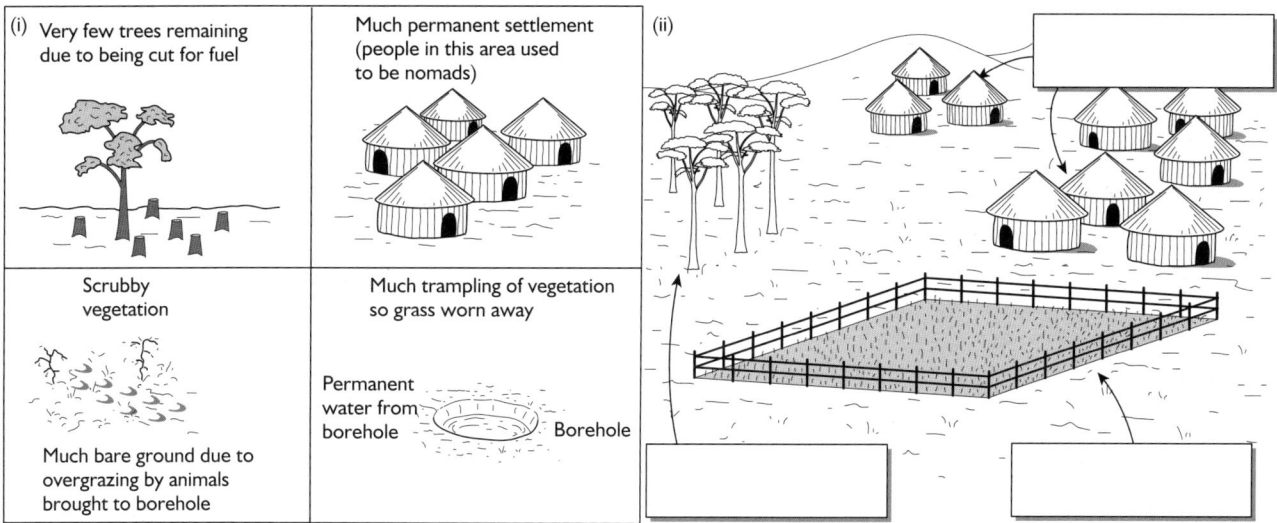

(i) Very few trees remaining due to being cut for fuel

Much permanent settlement (people in this area used to be nomads)

Scrubby vegetation

Much bare ground due to overgrazing by animals brought to borehole

Much trampling of vegetation so grass worn away

Permanent water from borehole

Borehole

A The settlement at Korr, after (i) the deeper borehole was dug and (ii) a sustainable development scheme launched.

3 Divide into groups of at least four for a role play exercise, which involves a discussion between two sets of people with different views on the future development of Korr. The discussion involves:

 • the local settled population who appreciate the services provided and do not want to move away

 • the local leaders, who see how nucleated settlement has damaged the environment and want to practise sustainable development in the area.

 a) In your group, write two lists of points representing the views of both sets of people.

 b) As a group, decide which set of people has the stronger argument. Be prepared to put your ideas and arguments to the whole class.

1 Explain briefly why trade between countries is necessary.   *(3)*

2 Define the following terms:

  a) *primary goods*

  b) *secondary goods*

  c) *imports*

  d) *exports*

  e) *trade balance*

  f) *transnational company (TNC).*   *(6)*

3 Study Figure A, which shows how the imports and exports of four selected countries are divided into primary (raw materials), secondary (manufactured goods) and other categories.

  a) Which two countries are MEDCs and which two are LEDCs?   *(4)*

  b) In which type of country do primary goods dominate trade?   *(1)*

  c) In which type of country do secondary goods dominate trade?   *(2)*

  d) What are the disadvantages to a country of relying heavily on the export of raw materials?   *(3)*

  e) How could such a country benefit more from its raw materials?   *(4)*

4 a) What type of graph is shown in Figure A?   *(2)*

  b) Why is this type of graph particularly useful for showing this type of trading data?   *(2)*

  c) Figure 11.11 on page 186 of the pupil book and the graph here are based on the same data for the same countries. However, they show the information in different ways. Which method do you prefer? Give your reasons. (Marks are given for your reasons, not your decisions, which are personal.)   *(4)*

5 Look at page 189 in the pupil book to answer the following questions.

  a) Which type of country dominates world trade today?   *(1)*

  b) Which continent is least involved in world trade?   *(1)*

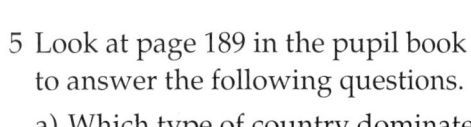

A Imports and exports for four selected countries.

  c) Which group of countries is most involved in world trade?   *(1)*

  d) What do the letters *GATT* and *WTO* stand for?   *(2)*

  e) Explain how GATT and WTO might improve the balance of world trade in the future.   *(5)*

1 What is meant by each of these terms:

  a) *aid* (2)
  b) *GNP?* (2)

2 Study Figure A and read the extract at the foot of the page.

  a) The extract refers to Official Development Assistance (ODA). To which category of aid does ODA belong? Figure 11.17 on page 190 of the pupil book will help you. (1)

  b) (i) Which two countries receive most aid as a percentage of their GNP? Use your atlas to help you name the countries you select. (2)

    (ii) How much aid did these countries receive as a percentage of their GNP? (2)

  c) (i) Name a country that received between 10 and 19% of its GNP as ODA. (1)

    (ii) Name a country that received between 0 and 4% of its GNP as ODA. (1)

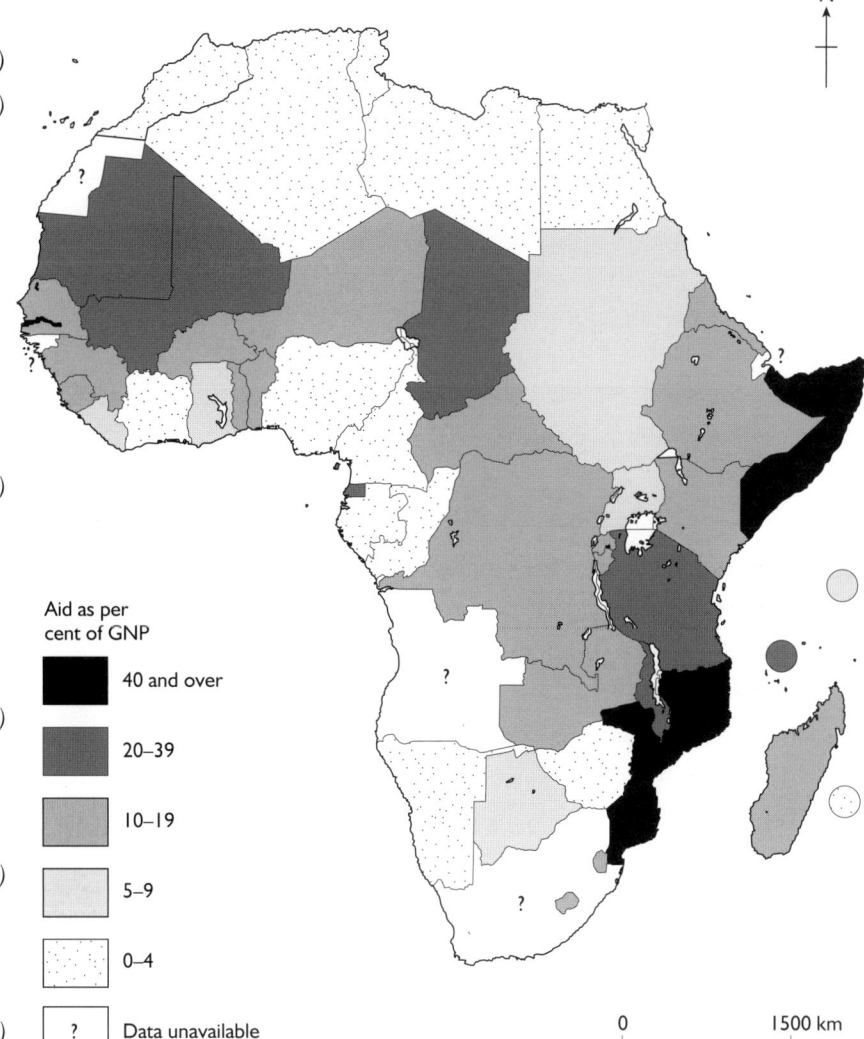

A Aid as a percentage of GNP in Africa.

  d) How much ODA does Africa as a whole receive annually from the industrialised countries? (1)

  e) (i) Discuss the problems experienced by the recipient countries as a result of this aid system. (6)

    (ii) Discuss the advantages and disadvantages for the donor countries of giving such aid. (The Case Study Extra on page 130 will help you.) (7)

Many African countries are dependent on large amounts of Official Development Assistance (ODA) from industrial countries. Africa as a whole receives over US$20 000 million in ODA every year. Egypt alone receives around US$1600 million. The Gambia receives ODA equivalent to over half its GNP and Mozambique and Somalia each receive over 40%.

Such a degree of dependence is alarming, even if ODA were always given by wealthier countries feeling a moral responsibility to help the poor. However, the reality is that motives for giving ODA are often not simply charitable. Aid can come in the form of armaments, which have more to do with supporting a government in the poor nation that is acceptable to the rich nation than with a desire to relieve poverty.

From **The Atlas of African Affairs (Second Edition)** by I. Griffiths.

Sponsoring Children in LEDCs

The Sponsorship System

You may have heard of this type of aid to developing countries. Indeed, your family or friends may already sponsor a child, or an elderly person, in this way. This type of aid is classified as voluntary (see Figure 11.17 on page 190 of the pupil book). Sponsorship fits in with most of the details in this table:

- it is not tied to any political or trading situation

- it encourages low-cost self-help schemes

- it often reaches poorer people in more remote areas

- the projects are dependent on the charity's ability to collect money and so annual support income can be uncertain.

In addition, it does not deal with emergencies. Its main aim is to improve the opportunities of individual children, as well as the quality of life of their families and communities, through low and intermediate technology development schemes.

People in MEDCs usually get to know about these support schemes because the charities concerned place advertisements in national newspapers or send appeals directly through the post. Charities today are very well organised. They take advantage of modern methods of targeting direct mail. For example, if you have donated to a Third World charity before, your name and address will be on that charity's mailing list. These lists are exchanged or sold between similar charities. Other charities may then approach you in the hope of a donation. The use of targeted mailing lists in this way cuts down a charity's costs and is more likely to increase donations, because the overall response is better than with random appeals. More of the charity's money is then available for aid work. The article opposite explains how the sponsor system actually works.

How it works

Ailsa Espie, 37, a nurse from Edinburgh, started sponsoring children six years ago in response to television adverts by the charity World Vision and sees it as a good way of supporting deprived communities.

At first, she was in touch with a child in Ecuador but when that project came to an end she started sponsoring an Indian boy, Srimante, who will be 12 in February. She donates £18 a month by direct debit.

She says: 'I started sponsoring around the time I started my own family. It seems like an easy way to make a contribution. Although I sponsor a child, my money doesn't go to the child, but to the community. Building wells in Ecuador benefited everyone.

'The children write as often as their sponsors write to them. I send a Christmas card and a birthday card.' She says she is also allowed to send modest gifts such as pencils and paper and receives a short report card about Srimante with details of things such as his favourite subject.

To start with, Ailsa was interested in sponsoring a girl, but when the first project she was supporting came to an end, she was asked if she would sponsor an Indian boy. She has found it a useful way of explaining development issues to her own three girls.

'I try and share with my own children that we are helping a boy in India and explain that going to school and having fresh water are luxuries for many people. One thing that goes through your mind is, the boy I'm sending to, how is he singled out from other deserving cases in the area?

'But I would like to think I was making a valuable contribution to the whole community in some small way. I see myself doing it for the foreseeable future.'

Recent New Publicity

In January 2003, the film *About Schmidt* was released in the UK, telling the story of a character who sponsors a child in this way. The film was actually based on a real sponsorship charity, Childreach, based in Warwick, Rhode Island, USA. The impact of the film company using Childreach is likely to be huge. Starring Jack Nicholson, the film will be seen by millions of people in the USA and the UK, so bringing this type of charity work to the attention of a much wider audience than normal and showing the benefits not just to the child concerned, but also to the donor.

'You can change a child's life,' says Childreach in a press statement about the film, 'maybe even your own.'

From **Glasgow Herald** *2 January 2003.*

UK Sponsorship Charities

Two examples of similar charity organisations in the UK are ActionAid and Families for Children. The first is larger and relatively well known, the second considerably smaller.

ActionAid

This charity runs a sponsorship programme and also ventures into education and awareness of Third World issues in UK schools. For example, their education officer produces a range of educational materials and regularly contributes to geographical publications such as *GeoActive* and *Geofile*. Begun in 1972, by 2001 ActionAid had 83 500 sponsors, as well as other supporters taking part in their monthly lottery, another fund-raising venture. They often advertise in newspapers for new sponsors, highlighting an individual child who 'needs your help'. Monthly sponsorship costs about £15. Ninety per cent of this goes directly overseas and 10% is needed for costs in the UK, including new fund raising. In the receiving region, 70% of the money is put into a central fund for various development projects within the child's village or wider community, so benefiting all. Villagers agree the nature of the projects at the outset. Twenty per cent pays for wider anti-poverty initiatives and the remaining 10% is for administration in the receiving country.

Some children are genuinely helped. The scheme allows them to go on to secondary education, which they otherwise might not have been able to do due to poverty and their family needing them to work and earn money from a relatively early age. College education opportunities are then also opened up for a few.

Families for Children (FFC)

This smaller charity was set up by British Airways cabin crew. Whilst on stopovers between long haul flights, some chose to use their spare time to help disadvantaged children. The result was the setting up of three orphanages, one in India, one in Somalia (which has since closed), and the third and most successful in Dacca, the capital of Bangladesh.

After some years, the Dacca orphanage moved from the city to a new site outside. It is on land about 10 m above sea level, which is high for Bangladesh as most of the country is at or only just above sea level, and therefore safe from flooding. Over 200 children live there at any time, from tiny babies to young adults. Not all are actually orphans; in some cases their parents simply cannot look after them, usually when the main breadwinner is lost, as when the father dies or is disabled.

The children are cared for by a number of local employees, as well as visitors from the UK who volunteer for a period of time. Some current British Airways crew, as well as some past members who were in at the beginning of the project, also still visit. The children are organised into small groups of mixed ages which aim to function as 'family' or 'household' units. Parents are also encouraged to visit. Some children are brought up throughout their whole childhood at the orphanage, but others only stay for a limited time until their family can get back on its feet again.

Each child in the orphanage has a sponsor in the UK. The cost to the sponsor is low and was about £5 per month in 2003. Each sponsorship goes solely to support an individual child, not other community projects, and this is one important difference between FFC and ActionAid.

In both charities' systems, the sponsor is kept in touch with the child in return for their regular donation. This involves being sent a photograph, which is updated regularly as the child grows and changes, school reports (usually twice a year), newsletters and sometimes letters from the child. Sponsors are encouraged to take an interest, but presents are to be kept small – hair ribbons, pencils and so on, so as not to differentiate too much between children. The sponsor is told not to give their UK contact details to the child to prevent any awkwardness later on. Letters are always exchanged through the organisation itself.

A A child in the FFC sponsorship scheme.

Problems with the Sponsoring System

The system does not always work well. For example, one girl from South India, handicapped by polio and sponsored by ActionAid, was withdrawn from the scheme by her father, who believed that after a certain age his daughter's place was in the home learning domestic skills, rather than in school. It could be argued that as a disabled person she would be less likely to get married and therefore had even more need than others to learn to support herself, but this example shows how the charities have to work within the cultural and social situation, which is not always easy. However, other children, and perhaps more often boys, do have greater chances in life as a consequence of such schemes.

Sometimes sponsors are prepared to do more for their children, but the system may prevent this. One girl in the FFC orphanage near Dacca wanted to be a nurse and her sponsor had enquired about the cost of funding this. By the time the enquiry was dealt with, the girl had already been placed in a textile apprenticeship, so the opportunity was wasted. Nevertheless, this girl had had much greater opportunities through FFC than she would have had without the charity.

Some charities and agencies, such as Christian Aid and Oxfam, deliberately choose not to run sponsorship schemes, preferring to work with whole communities. They feel that helping individual children can often cause inequality between brothers and sisters or between friends. Choosing which children receive the help can also be a problem, although this is usually done simply on the basis of need.

Child sponsorship has proved to be an important source of fund raising for many charities, especially for ActionAid and FFC, who are almost totally reliant upon it. It will be interesting to see whether the extra publicity this type of aid gets from the film *About Schmidt* increases people's response significantly.

Exam Practice Questions

1 Name one type of aid and give two of its advantages and two of its disadvantages. *(5)*

2 Imagine that you live in a village in the Third World that has been chosen by a sponsorship charity. Would you be in favour of such a scheme or not? Explain your reasons.

It does not matter if you decide whether you are in favour or not in favour, because marks are not given for the actual decision you make. Instead, marks are given for the reasons you give to support your decision, as well as for the quality of the arguments you construct. *(9)*

Sponsoring Children in LEDCs

These model answers are designed to show you how to get full marks for the longer question in the Case Study Extra for this section. Read the question carefully and write your own answer. Then read the model answer and the examiner's notes to see what he or she is looking for when awarding the marks. Decide how many marks you think the examiner would give your answer. Decide how to change your answer to this particular question to increase your marks. What will you do differently next time you answer a similar question?

2 Imagine that you live in a village in the Third World that has been chosen by a sponsorship charity. Would you be in favour of such a scheme or not? Explain your reasons. *(9)*

Version A The village accepts the charity's offer

If the charity invests money in the village the standard of living should increase, e.g. the children's education will be more secure and more will reach secondary education.

In most LEDC countries today, parents must pay a small amount for each child's schooling and for a poor family with several children this can be a severe burden. Most families only have spare money if they can produce a surplus, which can then be sold for profit. Also, going to school incurs other expenses, such as uniform, books, etc. If a charity sponsors a child these costs will be paid and the parents will then be able to give their other children a better education. Education is very important in an LEDC as it is seen as the only real way out of poverty.

Some charities fund whole village development so it may be possible to increase earning for the parents and siblings, leading to a better standard of living. Health care and other important services are also often a part of such schemes, so again the whole community benefits. In choosing some children for sponsorship others are inevitably left out, but if the poorest are chosen that would seem fair, and in any case the whole community does benefit from the larger-scale projects.

Version B The village does not accept the charity's offer

When a sponsorship charity comes into a village only some children, and therefore only some families, are chosen to take part. This means that others are very definitely left out. Resentment is likely to follow. Friends, both children and parents, may fall out as a consequence. Within a family, if one child is chosen over the others, then problems may also occur. Parents are put under pressure to provide exactly the same opportunities for the other children in the family, which may not be financially possible.

The charity that comes into the community may have fixed ideas concerning the type of development project it would like to see put in place. Arguments may therefore occur between members of the village. Some may feel they are being unfairly treated because they take a greater or lesser part in such projects.

Cultural perceptions vary between villagers and charity workers. For example, villagers may feel the right place for their daughters is at home doing domestic tasks, whilst the charity may encourage their education and work outside the home.

Improve Your Mark!

Level 1 (1–3 marks): *Too few reasons are given to support the decision made, no connections are made between the ideas and the answer is more like a list of statements.*

Level 2 (4–6 marks): *More reasons are given and some connections are made.*

Level 3 (7–9 marks): *Here a wide selection of reasons is given and many connections are made.*

CHAPTER 12 considers the topic of weather and climate in four main sections: Weather and climate, including air masses and factors affecting temperature, Rainfall, Depressions and Anticyclones and Weather maps. Case Study 12 considers seasonal and regional weather patterns.

Weather and Climate   📖 200 – 201

Pupils are given basic facts and definitions, which must be understood before they begin their study of weather patterns. Factors affecting temperature are considered for the British Isles, with maps showing isotherms and the influence of altitude on the duration of snow cover in winter. An understanding of air masses is crucial to pupils' grasp of weather systems. Britain's day-to-day weather, and its great variability, is the product of these systems. Figure 12.1 on page 200 relates the air masses and the directions from which they approach Britain to the weather they bring.

Activity Sheet 12.1 considers some general features of British weather – temperature, precipitation, growing season and hours of sunshine. Links are provided to other main GCSE and Standard Grade topics, including agriculture and recreation/ tourism. Question 4 asks pupils to explain the location of these types of activities in terms of weather characteristics.

Activity Sheet 12.2 concentrates on the factors affecting temperature. These are summarised on a map of the British Isles, which pupils must interpret to answer the questions. The influences of coastal or inland location and latitude are emphasised.

Rainfall   📖 202 – 203

With the help of diagrams, this section explains the processes involved in the formation of the three types of rainfall experienced in the British Isles. Pupils' understanding of frontal rainfall is then used later on pages 204–205 to develop their understanding of depressions.

Activity Sheet 12.3 gives pupils practice in explaining the differences between the three types of rainfall. They are asked to work on various aspects of all three in both written work and diagrams.

Depressions   📖 204 – 205

Depressions must be classed as one of the most difficult weather topics for pupils at this level and for this reason the subject has been included in Activity Sheets 12.3 and 12.4, to give extra practice. Figure 12.11 on page 204 relates weather conditions to a cross section through a depression. Figures 12.12 and 12.13 on page 205 introduce and link satellite weather images and synoptic charts.

Activity Sheet 12.4 concentrates on the interpretation of synoptic charts and station models, relating them to the relevant sectors of a depression.

Anticyclones and Weather Maps   📖 206

The synoptic charts and related satellite images on page 205 of the pupil book can be interpreted in various ways to aid pupils' understanding.

Activity Sheet 12.5 is directed primarily at high ability pupils. It diverges somewhat from the text in the pupil book, moving on to weather forecasting. Forecasts from newspapers, and a consideration of the equipment with which they are obtained, are included. The Meteorological Office website is given and other sites worth visiting are:
www.bbc.uk/weather/
www.csv.warwick.ac.uk/~esuvf/weather.html
www.cwp.co.uk/weather/

Case Study 12 Seasonal and Regional Weather Patterns   📖 207

The British Isles is divided up into four regions (north-west, north-east, south-east and south-west), which are linked to their climate graphs with explanations, providing a useful summary.

Case Study Extra Britain's Changing Weather and the Consequences

This case study goes into some depth about the weather pattern that developed to produce some of the worst floods in the year 2000. Uckfield and Lewes in East Sussex suffered particularly badly and their circumstances are explored. Possible future weather changes for the British Isles are also discussed.

Study the climatic maps Figures A and B and Figures 12.3–12.5 on page 201 of the pupil book.

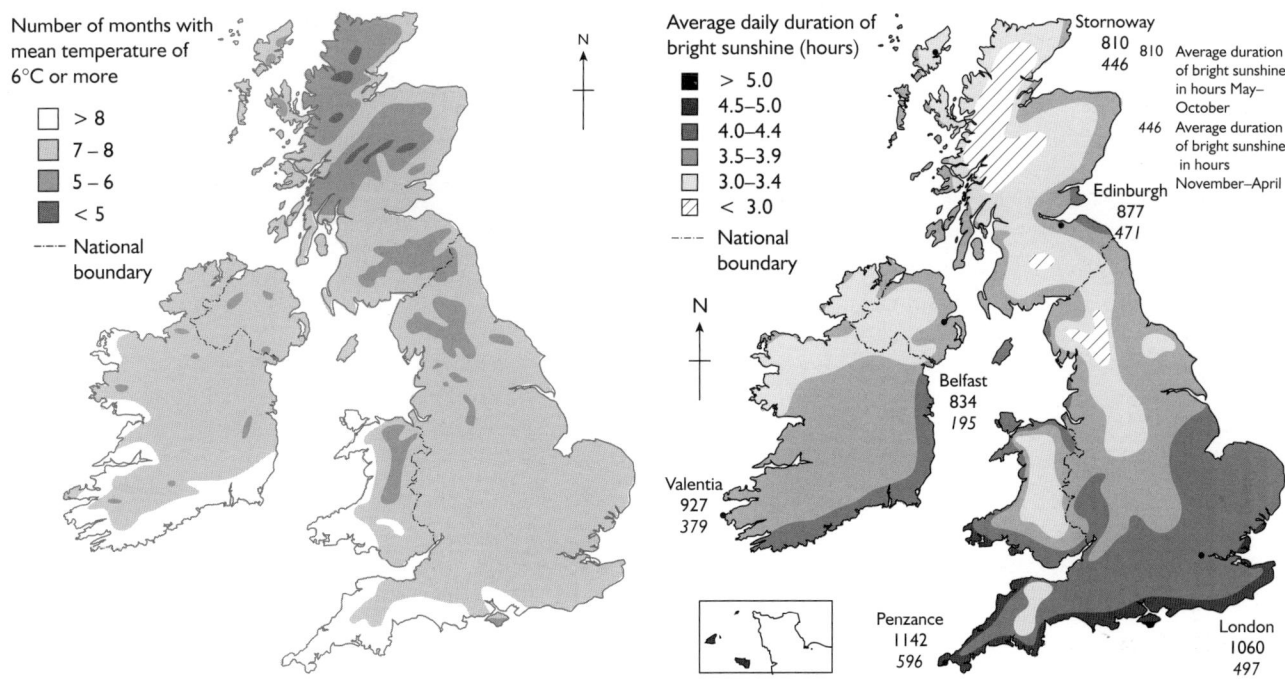

A Length of the growing season in the British Isles.

B Number of hours of sunshine in the British Isles.

1 What is meant by the term *growing season*? (1)

2 An isoline is a line on a map joining all the places of equal value in a particular respect.
   A contour on a map is therefore one example of an isoline, because it joins all the places at
   equal height above sea level. Define the following:
   a) *isotherm* (see Figures 12.3 and 12.4 on page 201 in the pupil book) (1)
   b) *isohyet* (see Figure 12.6 on page 202 in the pupil book) (1)

3 Referring to Figures A and B, answer the following questions.
   a) Where is the longest growing season? (3)
   b) Which areas have the most hours of sunshine? (3)
   c) What is the link between the two? (2)
   d) What other factors might be influencing the patterns of:
      (i) the length of the growing season (4)
      (ii) the number of hours of sunshine? (4)

4 Climate is a most important factor in determining the location of some human economic
   activities.
   a) (i) Name one area in the UK that is important for agriculture. What type of farming is
           important there? (2)
      (ii) Explain how the climate of your chosen area benefits the type of farming found
           there. (Hint: mention temperature, precipitation, sunshine and growing season.) (6)
   b) (i) Name one area in the UK that is important for recreation and tourism. (1)
      (ii) Explain how the climate has encouraged the type of recreation and tourism in this
           location. (4)

Figure A shows the factors affecting the temperature of the British Isles. You will need to use it and the information on page 201 of the pupil book to answer the following questions.

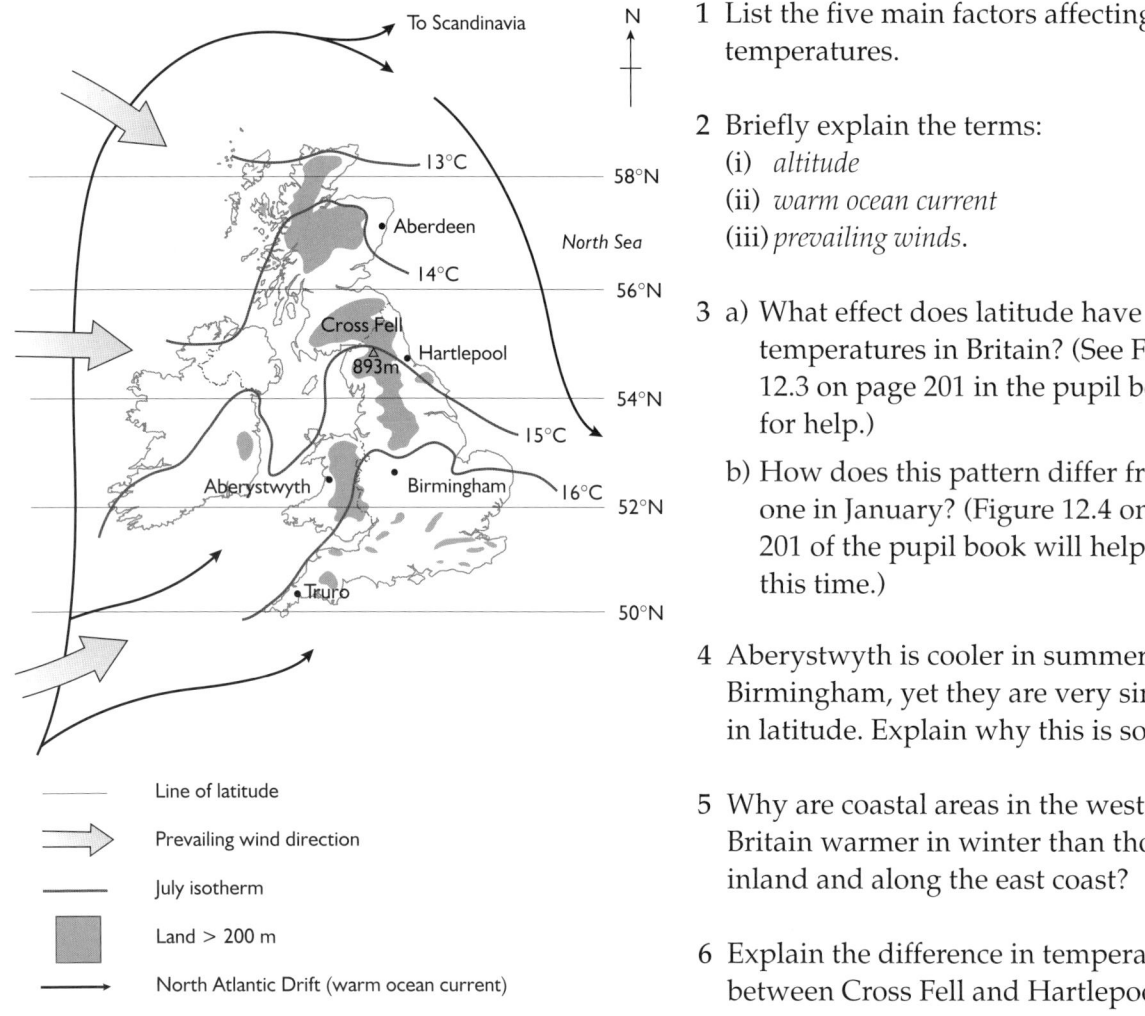

A Factors affecting the temperature of the British Isles.

1 List the five main factors affecting temperatures. (5)

2 Briefly explain the terms:
(i) *altitude*
(ii) *warm ocean current*
(iii) *prevailing winds.* (3)

3 a) What effect does latitude have on July temperatures in Britain? (See Figure 12.3 on page 201 in the pupil book for help.) (3)

b) How does this pattern differ from the one in January? (Figure 12.4 on page 201 of the pupil book will help you this time.) (3)

4 Aberystwyth is cooler in summer than Birmingham, yet they are very similar in latitude. Explain why this is so. (2)

5 Why are coastal areas in the west of Britain warmer in winter than those inland and along the east coast? (2)

6 Explain the difference in temperature between Cross Fell and Hartlepool. (2)

7 a) What is the difference in July temperature between Aberdeen and Truro? (1)

   b) Explain, in as much detail as possible, the reasons for this temperature difference. (5)

8 a) From which direction are the prevailing winds that affect most of Britain? (1)

   b) Over what surface have these winds blown before they reach Britain? (1)

   c) What effects do these winds have on temperature:

   (i) in summer (2)

   (ii) in winter? (2)

You should have used every piece of information provided on Figure A to answer Questions 4 to 8. If you have not done so, carefully check your work.

Figure A shows the position of a typical depression passing over the British Isles. The warm and cold fronts, as well as the areas experiencing rainfall, are marked.

1 Define the terms:
   (i) *depression*
   (ii) *warm front*
   (iii) *cold front*. (3)

2 a) In which parts of the British Isles is it raining in Figure A? Your atlas will help you to identify areas by name. (3)

   b) How do the positions of the depression and the fronts explain the locations of the areas of rainfall? (3)

   c) How can the small patches of rain to the west of Ireland and Scotland be explained? (2)

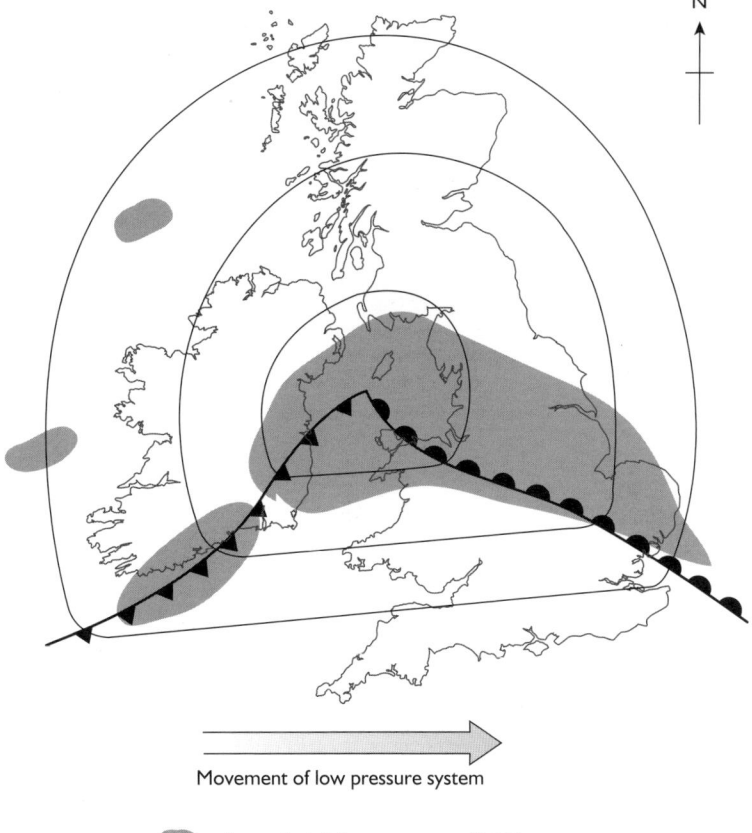

Movement of low pressure system

🌫 Area of rainfall   ▼▼ Cold front
🔴🔴 Warm front   —— Isobars

A Rainfall associated with low pressure systems in the British Isles.

3 a) Which type of rainfall is falling in the shaded areas on the map? (1)

   b) Describe the formation of this type of rainfall. (5)

   c) There are two other types of rainfall. Name them. (2)

   d) Figure B shows one type of rainfall in the north of England. Which one does it show? (1)

   e) Add labels to the blank boxes to explain this rainfall formation. (3)

   f) Draw diagrams to show how the third type of rainfall occurs. (Figure 12.10 on page 203 of the pupil book will help you.) (6)

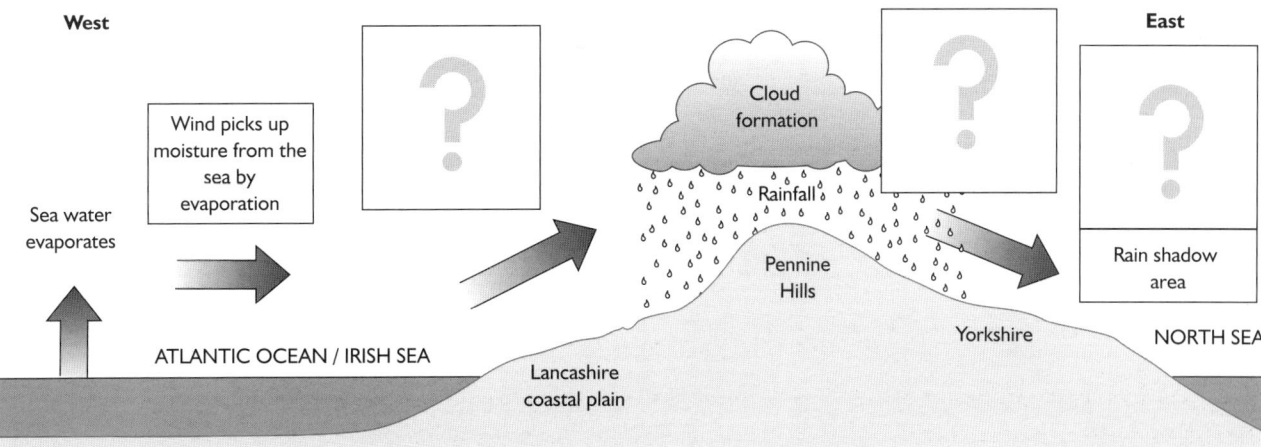

B The formation of rain in the north of England.

1 a) Figure A(i) and (ii) shows two weather symbols that would typically appear on a synoptic chart (weather map). Complete the table below to show the weather represented by each station model. (10)

(i)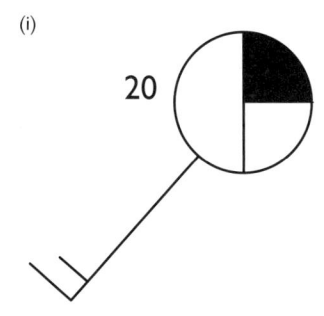

(ii)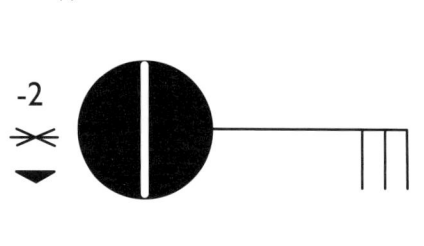

(iii)

A Station models.

Weather	A	B
Cloud cover		
Precipitation		
Temperature		
Wind direction		
Wind speed		

b) Station model C, Figure A(iii), is made up. It could not describe a real weather situation because it shows weather conditions that could not possibly occur together. Write down three of these conditions. (3)

c) Draw a station model to show the following weather conditions: (i) 0 °C, (ii) 6 oktas of cloud cover, (iii) a northerly wind of 10 knots and (iv) drizzle. (5)

d) Draw a station model like model A (iii) showing a contradictory situation. Test it out on others in your class to see whether they can see the problem. (5)

2 Figure B shows weather systems A and B, which affect the British Isles.

a) Name the two systems. (2)

b) Name sectors X and Y. (2)

c) Sector X has very different weather conditions from Sector Y. Describe the conditions and explain in detail why this is so. (6)

d) Describe the weather conditions in the high pressure system. (4)

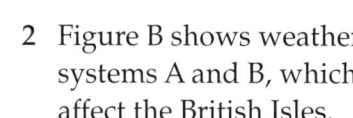

B Typical winter systems over the British Isles at noon.

1 Weather forecasts are reported to the public by various types of media. Write down three ways in which this information reaches us.

(3)

2 The weather map in Figure A is from one of the national newspapers. It uses different symbols from those you have already become familiar with on the synoptic charts.

a) Compare the weather experienced in London and in Aberdeen on 23 November 2002. (3)

b) What are the differences between the newspaper symbols and the ones on synoptic charts? (2)

c) Which set of symbols do you prefer using? Give your reasons.

(3)

3 The map in the article below shows the locations of some of the weather stations that record the weather around the British Isles. They use radiosonde balloons to take measurements. From these readings, weather forecasters use computers to predict the weather we are likely to experience.

A The weather forecast for the British Isles on 23 November 2002.

a) Why are the weather stations located in these positions? (2)

b) Why are radiosonde balloons so useful to weather forecasters? (4)

Radiosondes

The atmosphere is three-dimensional and a knowledge of what is happening above us is of central importance to weather forecasting.

One of the simplest and most effective methods of taking 'soundings' through the atmosphere is to release a radiosonde, a package of instruments carried aloft by a balloon. The sonde is tracked by radar, which gives wind speed and direction, while a transmitter continuously reports details of temperature, humidity and pressure to computers at ground stations.

The height reached by some sondes is over 30 km, although 20 km is the more usual height at which the balloon bursts and the instrument package parachutes back to earth.

The Meteorological Office, as part of the global radiosonde network, maintains eight upper air stations in the UK. Sondes are released from these, and from the weather ships at sea, as a daily routine.

Weather stations in the British Isles.

4 Using your school library and the Internet (for example, the Meteorological Office website www.metoffice.gov.uk/), find out about some of the equipment used for gathering weather data. Exchange your information with others in your class.

Britain's Changing Weather and the Consequences

Changing Weather

In Britain, flooding occurs on average nearly twice as frequently as it did a century ago. It has been predicted that over the next century, the risk of floods will be ten times as great. The increasingly unpredictable risk of floods is one of the first effects of global warming, which is a result of the emission into the atmosphere of higher levels of greenhouse gases, created by man's economic activity.

The costs of flooding are huge. Between 1998 and 2000, 25 people died as a result of flooding in England and Wales. Five million people live in low-lying areas that are prone to flooding. The value of property and land at risk has been estimated at around £220 billion. The Environment Agency has advised against new developments in flood plains, but at the same time the four or five million new homes that are needed in Britain before 2016 have to be built somewhere.

The autumns of 2000 and 2001 both saw considerably higher than normal rainfall, with depressions following so quickly on the heels of each other that the rainfall from one did not have time to drain away before it was added to from the next depression. In June 2000, floods hit Yorkshire and County Durham. In September it was Hampshire's turn, then East Sussex and Kent suffered the worst inundations of the year in October (see Figure A).

The East Sussex and Kent Floods in October 2000

Over 150 mm of rainfall fell on East Sussex and Kent in a twelve-hour period, causing the Rivers Uck and Ouse to turn the towns of Uckfield and Lewes, respectively, into rescue zones. Homes, shops, industrial units and abandoned cars were awash with muddy water. Lifeboat crews and a coastguard helicopter were called in to undertake inland rescues.

A Flood locations in Britain in 2000.

Residents were evacuated to public buildings. The River Medway in Kent caused serious damage to a number of villages, such as Yalding, while also threatening towns like Maidstone and Tonbridge. The national news reported these events as the worst flooding in decades.

The weather sequence

The Atlantic map on Figure B(i) on page 141 shows a complex series of seven fronts associated with three different depressions affecting the British Isles or on the way towards or just passed them. Several days of almost continuous rain occurred, so that all the ground became saturated. In the midst of this, the twelve-hour period of torrential rain caused large-scale surface runoff and consequent floods.

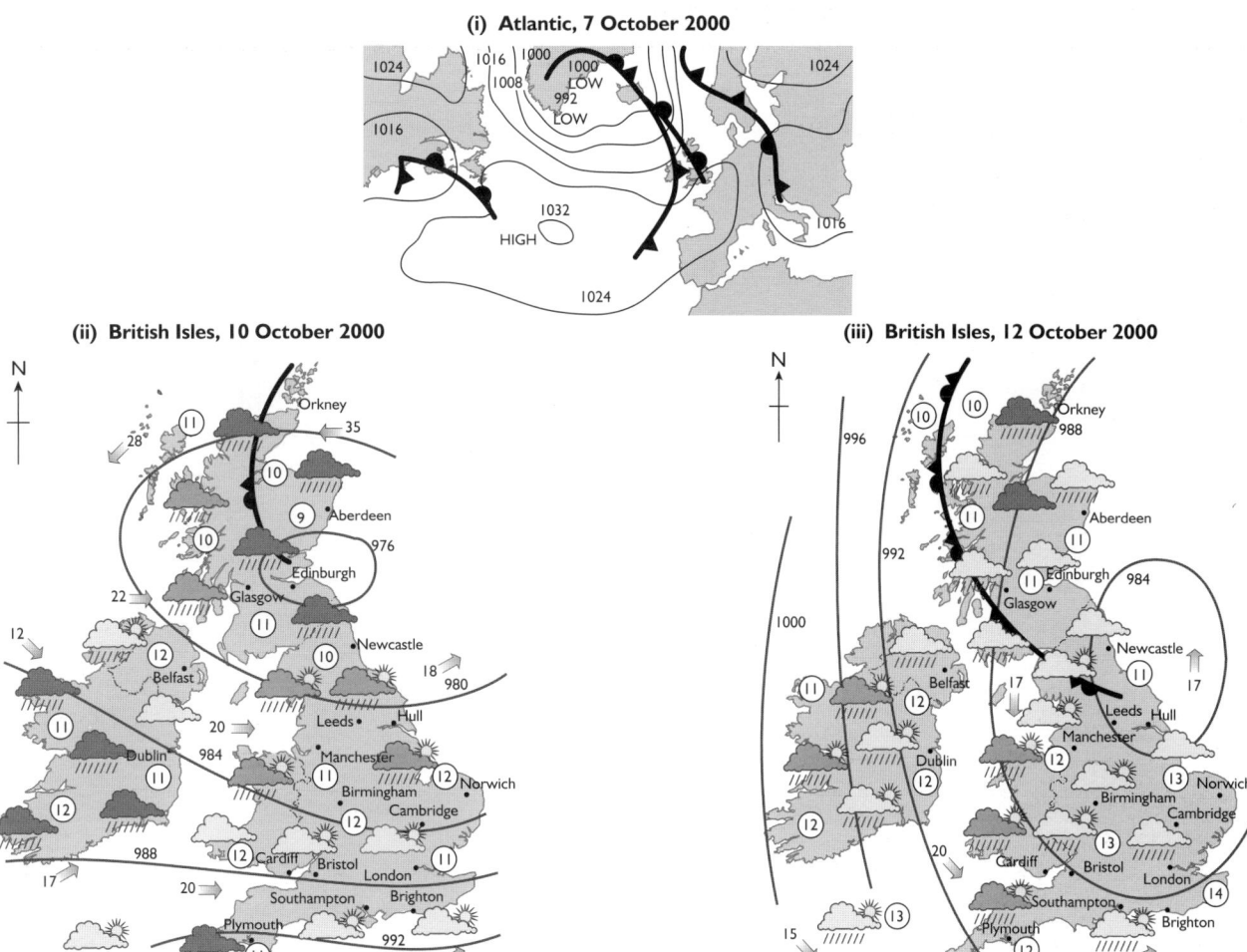

B The weather leading up to the Uckfield and Lewes floods in 2002 in (i) the Atlantic on 7 October and over the British Isles on (ii) 10 October and (iii) 12 October.

The weather map for 10 October in Figure B(ii) shows a deep depression (down to about 976 mb) centred over Edinburgh. The next day, that depression deepened even more to 966 mb and heavy cloud and rainfall moved over south-east England from Ireland and the Midlands. By 12 October, shown in Figure B(iii), and the day of the worst flooding, skies were clearing. This delay between the storm and the flood is explained by lag time, the period between peak rainfall and peak discharge. The height of the flood was during the early hours of the morning, but it was not until daylight that the true extent of the devastation was evident.

Britain's Future Weather Patterns

Two main theories have been put forward for predicting the results of global warming on Britain's climate. On one hand we might have a warmer climate, similar to that of the Mediterranean.

A Mediterranean climate

The chance of having an extremely cold winter would decrease. Milder, wetter winters would be the norm and extreme weather in the form of storms like those of October 1987 and January 1990 would probably occur more frequently. A more Mediterranean climate would lead to substantial changes in our ecosystems, agriculture and tourism. Ecosystems are very sensitive to changes in moisture. A Mediterranean-type summer would bring hotter, drier conditions or drought, so many British plants would be completely unable to survive. Species of typically British plants and animals would have to migrate northwards or they would become extinct. Continental species, such as might be seen in southern Europe, might then replace the British ones.

In a Mediterranean climate, crop productivity would also change. The growing season would be almost twelve months long. Wheat can be grown in a Mediterranean climate, but barley

The New Wider World (Second Edition): Teacher's Resource Book
Neil Punnett and Alison Rae © Nelson Thornes 2003

thrives only in cooler conditions. Vineyards might become more common. Today vines really only grow in Kent and Sussex, and these areas are on the very northern margins of possible production. However, such a climate change would bring about a repetition of the situation in Britain in Roman times, when such crops were grown as far north as the Midlands. New crops like olives might also be introduced. Gardens would be similarly affected, with traditional species failing, new ones being introduced and hardly anyone buying a greenhouse!

Tourism might be affected both positively and negatively. In Scotland, the skiing business in the Cairngorm Mountains would end, but coastal districts would become more attractive to holidaymakers looking for the sun, who may then choose not to pay for foreign travel.

Demand for energy would also change. Less energy would be needed for winter heating, although the demand for summer air conditioning would be likely to increase. Several power stations have coastal locations and these might be affected by a rise in sea level.

A cooler climate

The other main theory concerning the impact of global warming on the British climate is less attractive. Global temperature change might affect the positions of the ocean currents, so that the North Atlantic Drift would no longer move warmth from the Gulf of Mexico to British waters and we might therefore actually become colder in a warmer world!

Exam Practice Questions

1 a) On Figure B(i), label each front, using the following terms:

- warm front
- cold front
- occluded front.

You should have seven labels. (7)

b) Explain how a front can result in heavy rainfall. (4)

2 Using Figure B(ii) and (iii), describe the pattern of weather that occurred across the British Isles between 10 and 12 October 2000. Consider pressures, temperatures, precipitation, sunshine and wind speed and direction. (7)

3 Explain how certain weather patterns can result in flooding like that in Kent and East Sussex in October 2000. (4)

Britain's Changing Weather and the Consequences

These model answers are designed to show you how to get full marks for the longer questions in the Case Study Extra for this section. Read the question carefully and write your own answer. Then read the model answer and the examiner's notes to see what he or she is looking for when awarding the marks. Decide how many marks you think the examiner would give your answer. Decide how to change your answers to increase your marks. What will you do differently next time?

2 Using Figure B(ii) and (iii), describe the pattern of weather that occurred across the British Isles between 10 and 12 October 2000. Consider pressures, temperatures, precipitation, sunshine and wind speed and direction. *(7)*

> On 10 October 2000, a deep depression (about 976 mb) was sitting over the British Isles, centred approximately on Edinburgh. An occluded front lay over eastern, central and northern Scotland. Most of the British Isles were experiencing some rain, often heavy. Southern England and Northern Ireland had some sunny spells. All areas were mild, between 9 and 12°C. Most winds were westerlies, except in Scotland, where easterlies dominated. Wind speeds were high, especially off the coasts of north-east Scotland and Sussex, which was not surprising given the closeness of the isobars.
>
> By 12 October, the weather system had moved further south, so that it was centred on the east coast between Hull and Newcastle. The south and west still had rain, but the east was drier. Temperatures were a little higher, up to 14°C. Winds had eased as the pressure gradient was less steep.

> **Improve Your Mark!**
>
> Work through each feature of the weather, always pointing out any links. This question asks for description, so do not explain as well.
> **Level 1 (1–3 marks):** *A few points, not giving the picture across the whole country.*
> **Level 2 (4–5 marks):** *More description from the weather maps would be included and you would be starting to link the different features of the weather together.*
> **Level 3 (6–7 marks):** *For top answers describe the weather system in a logical sequence so it is clear how it is moving across the whole country.*

3 Explain how certain weather patterns can result in flooding like that in Kent and East Sussex in October 2000. *(4)*

> The various fronts on the map giving the Atlantic situation show just how much rainfall the British Isles received in the first half of October 2000. Very low pressure and strong winds could have led to stormy weather with spells of very heavy rain. The ground would have become saturated quickly so continued rainfall would easily have resulted in overland flow. River levels would therefore have risen exceptionally quickly, causing the severe flooding.

> **Improve Your Mark!**
>
> Refer to all the information given, as well as using your own knowledge. You must make clear links between the weather and its impact on the land surface. Comment on what led to the storm and that saturation of the ground and then flooding were bound to follow.
> **Level 1 (1–2 marks):** *You will get these marks if you only state facts from the resources provided.*
> **Level 2 (3–4 marks):** *You will get higher marks if you link the weather conditions to the floods.*

CHAPTER 13 presents eight sections and two case studies on world climate. Beginning with the atmospheric circulation that controls world climates, a selection of major climatic types is considered. Hurricanes are highlighted as a particular weather phenomenon of interest, as are the various issues surrounding global warming.

Atmospheric Circulation and World Climates
📖 210 – 211

The nature of world climate patterns cannot be explained without some background on the way in which the Earth's atmosphere operates and creates the airflows that are largely responsible for the conditions experienced over the globe. Figure 13.1 on page 210 provides a summary of this system as a basis for the study of global weather patterns.

Activity Sheet 13.1 looks at the world's pressure and wind systems, beginning with the three atmospheric cells. Pupils must label the world map with details of the pressure and wind systems.

Activity Sheet 13.2 relates to the section on factors affecting global temperature on page 211, with particular emphasis on the influence of ocean currents.

(**Activity Sheet 12.2** considers the factors affecting temperature in the British Isles and could therefore be used with either chapter.)

Equatorial and Tropical Continental Climates
📖 212 – 213

Rainforest climates are considered with regard to their annual and diurnal patterns. A tropical continental (interior) climate demonstrates wet and dry seasonality. Both of these climates provide a useful link with the biomes studied in Chapter 14.

(**Activity Sheet 14.4** also uses a climate graph for a savanna region, providing a link with the tropical continental (interior) climate here.)

The Monsoon, Mediterranean and Cold Climates
📖 214 – 215

Maps of the Indian subcontinent explain the monsoon weather patterns. Links could be made to Case Study 7 on rice cultivation on pages 112–113. Malta and Fairbanks, Alaska, are used as examples of Mediterranean and cold climates.

Activity Sheet 13.3 develops pupils' skills in constructing and interpreting climate graphs, as well as comparing world climates. Statistics are given for Moscow, a cool temperate interior climate.

Tropical Cyclones (Hurricanes)
📖 216 – 217

Although weather can cause several categories of disaster – floods, drought, blizzards, avalanches, etc. – hurricanes are simply so dramatic in their impact that they appeal to pupils in much the same way as volcanic eruptions and earthquakes. Hurricanes Mitch and Andrew are contrasted as MEDC and LEDC examples. The Case Study Extra on pages 151–153 of the *Teacher's Resource Book* also tackles hurricanes.

Global Warming
📖 218

The greenhouse effect is explained and the causes of global warming listed.

Effects of Global Warming
📖 219

Rising sea levels, increased flooding and changes to ecosystems are considered.

Activity Sheet 13.4 asks pupils to consider possible future scenarios. It might be useful to hold a class debate on these issues prior to completing this sheet.

Predicted Effects of Global Warming
📖 220

UK and world situations are considered, with detailed annotated maps of both.

Acid Rain
📖 221

This page provides illustrations of human impact on natural systems and processes, emphasising the European situation.

Activity Sheet 13.5 looks at the specific impacts of global warming in the UK, particularly on ecosystems, providing a link with Chapter 14 in the pupil book.

Case studies 13a and 13b Drought and Water Supply in the UK and in Developing Countries
📖 222 – 225

The contrast between MEDC and LEDC situations in this section provides a link with the development issues presented in Chapter 11 in the pupil book. The studies also illustrate the potential climate changes resulting from the global warming scenarios considered earlier in this chapter.

Case Study Extra Hurricane Zoë, 29 December 2002

While not a huge event in terms of the number of people and the area of land involved, this recent hurricane was significant in highlighting the problems that can be caused by remoteness during a natural disaster of any type.

1 Refer to the pupil book to help you explain how the following conditions are caused:

  a) low pressure *(2)*

  b) high pressure *(2)*

  c) wind. *(2)*

2 Over each hemisphere there are three atmospheric cells.

  Name the three cells and label Figure A to show where they are located. *(6)*

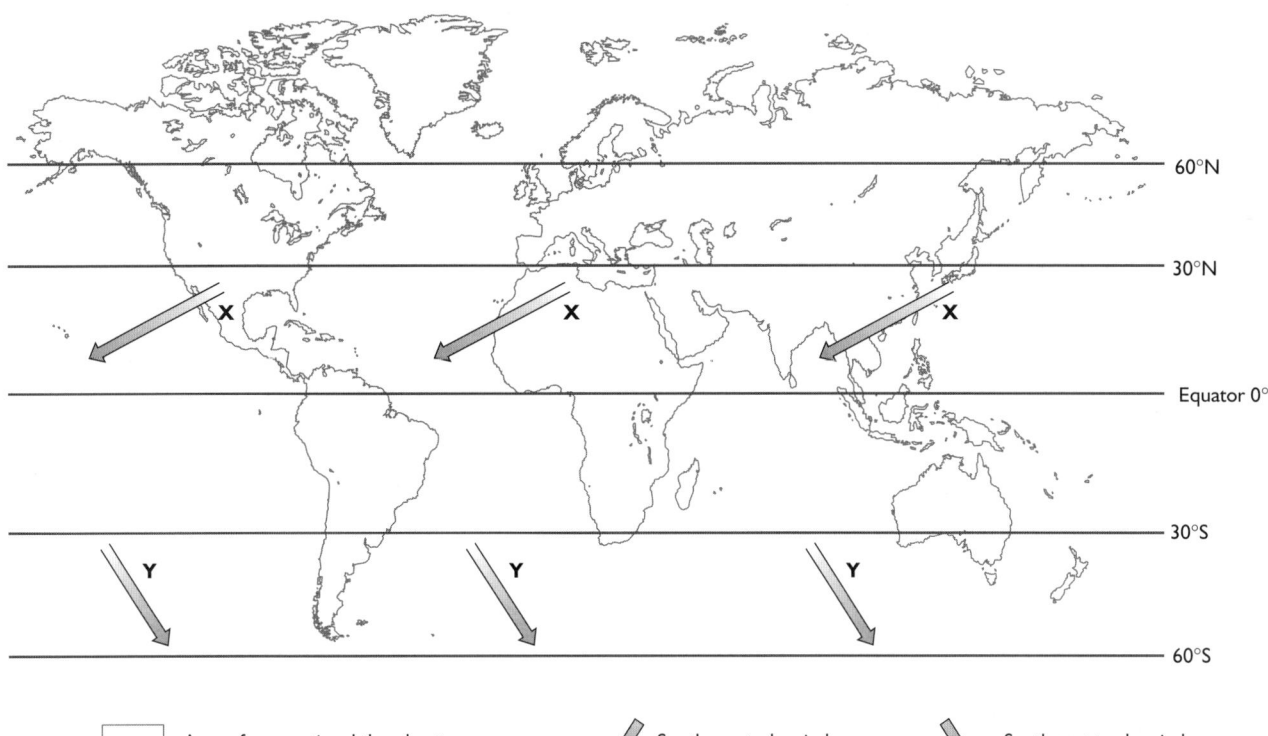

A Selected global winds.

3 Use Figure A to answer the questions below.

  a) On the map, shade in an area where you would expect to find convectional thunderstorms. Complete the key to identify the area. Write down one reason why such storms are found in your chosen region. *(3)*

  b) Two winds, X and Y, have been marked on the map. Which global winds do these arrows represent? *(2)*

  c) Using the style of arrows given in the key, mark the following on the map:

   (i) an area experiencing south-westerly winds *(1)*

   (ii) an area experiencing the south-east trade winds. *(1)*

4 a) What is meant by the term *season*? *(1)*

  b) When is the northern hemisphere at its hottest? *(1)*

  c) When is the southern hemisphere at its hottest? *(1)*

  d) Explain how the Earth's tilt and rotation affect our seasons. *(4)*

  e) Name a part of the globe that does not have seasons based on temperature. Explain why this is so. *(4)*

1 a) List five factors that affect temperature. (5)

  b) Explain briefly how each factor operates. (5)

2 Study Figure A, which shows the major ocean currents. Some currents are classified as 'warm' and others as 'cold'.

  a) Identify the main difference between warm and cold ocean currents.
    (Hint: think about direction of flow.) (2)

  b) Name and locate two warm and two cold ocean currents. (4)

  c) Describe the pattern of ocean current flows in the North Atlantic. (4)

  d) Describe the general global pattern of ocean current flows. (4)

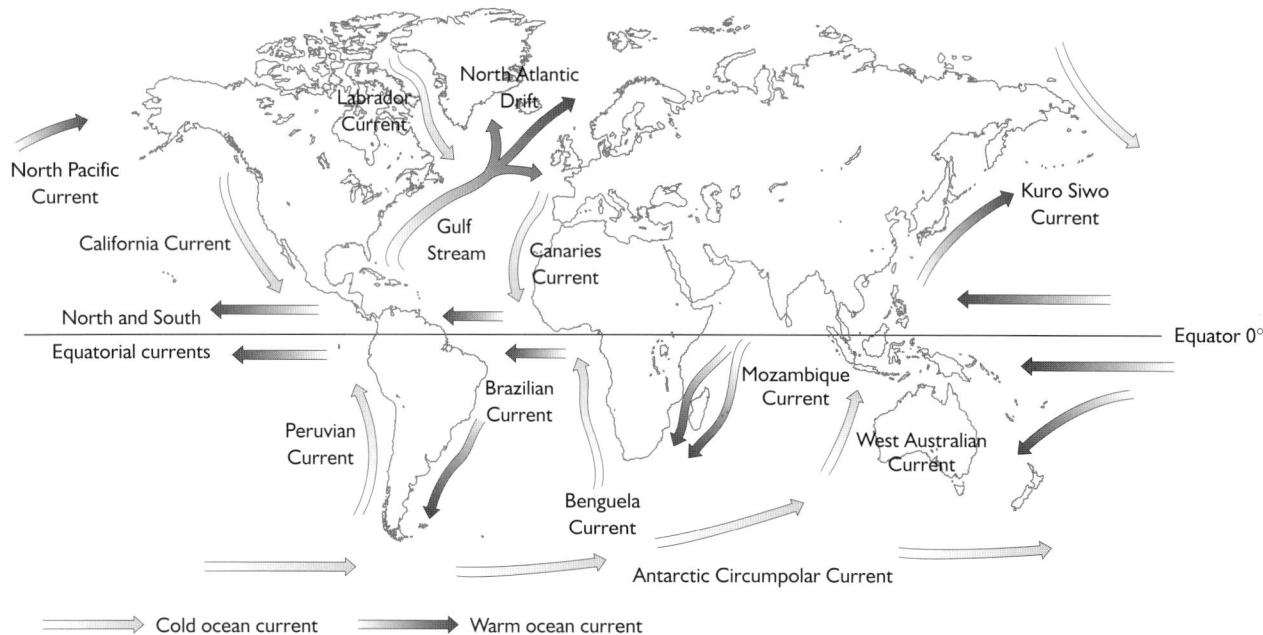

A Major world ocean currents.

3 Coastal areas of land are more affected by ocean currents than the centres of continents.

  a) Explain briefly why this is so. (2)

  b) Using the table below, list some of the influences ocean currents can have on the climate of an area. One line has been done for you. (6)

	Warm current	**Cold current**
Temperature		
Rainfall		
Fog	Fog uncommon	Fog occurs offshore
Icebergs		

4 Choose one ocean current for further study. Find out all you can about it in your library or on the Internet.

  a) Name your chosen current. Highlight it clearly on Figure A. (1)

  b) State whether it is a warm or cold current and which area of the world it affects. (2)

  c) Describe and explain the impact of the current on the land area adjacent to it. (5)

1 The climate statistics in Figure A show the characteristics for a major world climate. Look at the climate graphs for each of the five climate regions considered on pages 212–215 of the pupil book. In the same way, plot the statistics from Figure A onto the graph frame in Figure B. Use red for the temperature line and blue to shade the precipitation bars. *(12)*

Months	J	F	M	A	M	J	J	A	S	O	N	D
Temperature (°C)	-5	-4	0	8	15	21	24	22	15	7	2	-3
Precipitation (mm)	28	25	37	40	45	52	70	73	55	34	41	33

A Climate statistics.

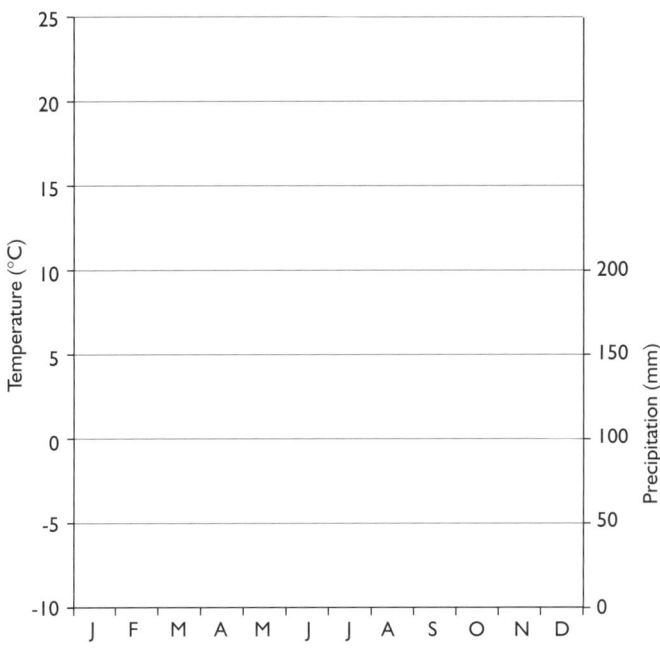

B Climate graph.

2 This question is to do with reading your graph.

   a) Why are the colours red and blue used for temperature and precipitation respectively in climate graphs? *(2)*

   b) What exactly is the difference in meaning between the terms *rainfall* and *precipitation*? Can they be used interchangeably? *(3)*

3 This question is to do with interpreting your graph.

   a) Describe the patterns of (i) temperature and (ii) precipitation shown by your graph. *(6)*

   b) Where in the world might this climate be found? *(2)*

   c) Using your knowledge of the factors affecting world climates, try to explain the reasons for this climate type being in this region. *(5)*

4 a) Compare the graphs in Figures 13.3 and 13.7 on pages 212 and 214 of the pupil book. *(6)*

   b) Give reasons for the differences you identify. *(6)*

1 Explain what is meant by each of the following terms:

a) *the greenhouse effect* (2)

b) *global warming* (2)

c) *emissions* (2)

d) *CFCs* (2)

2 Name four causes of global warming and explain how each one operates. (8)

3 The headlines below are not real – not yet, anyway! If they were ever published it would be because global warming had become more extreme than today, as is predicted. Explain the situation that might lead to each of these headlines being printed. (6)

Warmer weather brings threats of malaria and cholera

Global warming brings olive trees to Britain

World's climate changes more rapidly in last 100 years than in previous 10,000

4 Figure A(i)–(iii) is also making points regarding global warming. Try to explain the points being made. Refer to particular world locations in your answers. (6)

5 Research in your library or on the Internet to discover ways in which we can combat global warming trends. Try the Friends of the Earth website at www.foe.co.uk.

A What's happening with global warming?

1 Define the term *acid rain*. (2)

2 Is rainfall naturally acidic or alkaline? What is its pH? (2)

3 Name two manmade pollutants that lead to acid rain and identify their sources. (4)

4 a) List five ways in which acid rain affects physical and/or human environments. (5)

   b) Choose two of these effects and use them to explain the nature of the impact of acid rain in more detail. (6)

5 Study Figure A.

   a) Which areas in the UK have the highest and lowest levels of average acidity? (4)

   b) Which of the following factors is your answer to (a) related to:

   • wind direction

   • height above sea level (4)

   • amount of urbanisation?

   Give a reason for your answer.

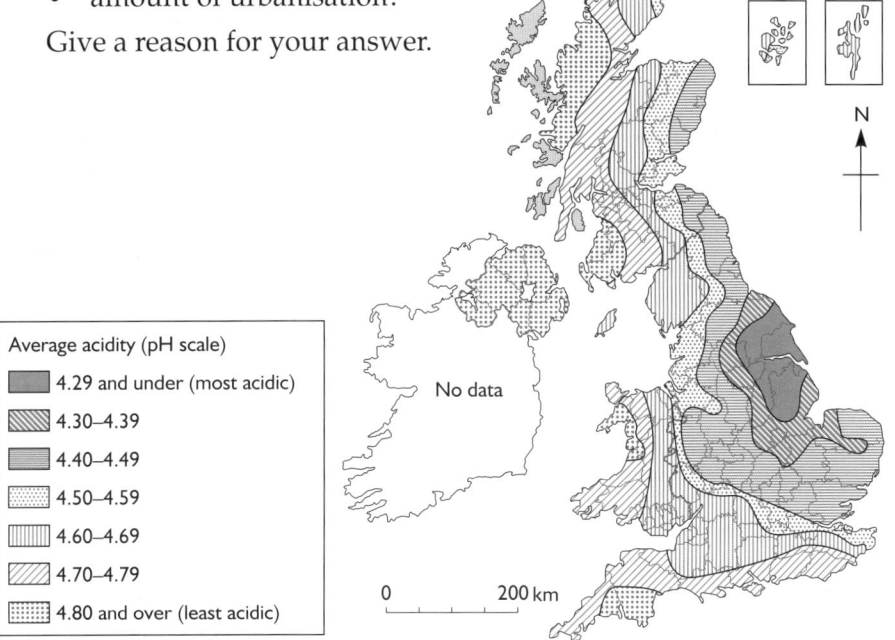
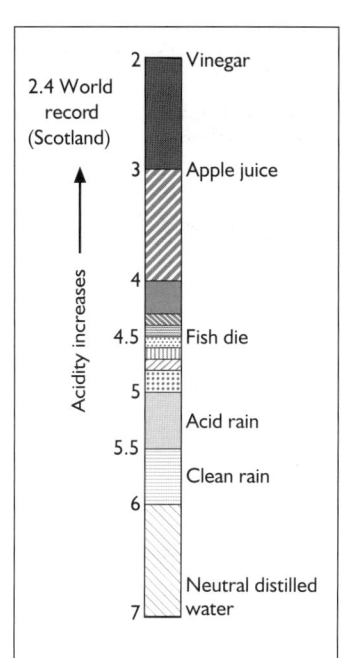

A Acidity of rainfall in the UK.

6 Read the extract on page 150, which gives more detail on acid rain pollution problems in part of Britain.

   a) What is an SSSI? (2)

   b) Name a region in which much of the land is categorised in this way. (1)

   c) How much land is actually affected in this region? (1)

   d) What specific impacts have there been in this region on:

   (i) fish (3)

   (ii) plants? (3)

   e) Which large mammal is judged to be at risk? (1)

   f) What are the likely future consequences for this region if the problems caused by acid rain are not tackled? (3)

Acid rain causes damage to British wildlife

The headwaters of the River Severn are so acidified by airborne pollution from distant power stations, industry and vehicles that they are unable to nurture the shoals of minnows and sizeable brown trout that once were abundant here. The upper River Severn has no fish …

Hill streams and lakes across northern England have acidified, too …

A report by English Nature, in collaboration with the Countryside Council for Wales and Scottish Natural Heritage, shows more than 430 000 hectares of land protected as Sites of Special Scientific Interest (SSSI) – our most important wildlife locations – in England, Scotland and Wales are suffering acidification damage. This represents almost a quarter of Britian's SSSI area.

In the most severely affected region, North Wales, 57% of the SSSI area – more that 48 000 hectares of land – is degraded or highly likely to have been degraded by a cocktail of sulphur dioxide and nitrogen oxides, which can, on occasion, transform pure rainwater into a brew more akin to vinegar.

As a result, many of the plants and animals on the rugged moors and bogs are in decline or have already disappeared. Conservationists are concerned for rare fish species such as powean and arctic char; for lake and stream populations of trout, salmon and minnows; for frogs; elusive otters; birds such as dippers, whose survival depends on healthy populations of water-living insects; and for the fragile crust of lichens and mosses that clothe the peat bogs over vast tracts of mountain and moor. It is a depressing list.

The largest amount of pollutant that will not cause chemical effects in the soil leading to long-term ecological change is known as the 'critical load'. Maps of soil critical loads produced by the Department of the Environment's Critical Loads Advisory Group show that more that 108 000 sq km, chiefly in Wales, Scotland, Cumbria, and northern Pennines, Dartmoor, Bodmin Moor and the New Forest, exceed their critical load. Their capacity to neutralise any more acid rain falling on them is already exhausted, so ecological change is unavoidable.

From **The Independent** *8 June 1992.*

Hurricane Zoë, 29 December 2002

A Small but Serious Disaster

Most of the hurricane case studies found in GCSE and Standard Grade materials concern the USA, Caribbean or Central America. Admittedly these areas do suffer badly, but they are also reported so widely because they affect either a developed country or its neighbours.

In this respect Hurricane Zoë, active right at the end of 2002, was different. Most of the damage it wreaked was on a few tiny Pacific islands with relatively few inhabitants. However, the disaster highlighted the fact that the very remoteness of these communities could cause severe problems (see Figure A). It took a rescue ship three days to even reach the stricken area with any sort of assistance. The first real reports in UK newspapers appeared on 2 January 2003, four days after the event itself, and by that time the rescue ships had not even reached their destination. A few aerial and satellite photographs were available from which the scale of the disaster could be estimated.

Just as with floods, we can expect hurricanes to be of very different degrees of severity and to occur on differing scales of regularity. Hurricane Zoë was assessed as the 100-year

hurricane, a very serious one, given that strong hurricanes occur in the same part of the Pacific every 20 years. The fact that major events are experienced every 20 years does mean that the people affected have developed some strategies for protecting themselves.

Battered islands show ravages of storm
Photographer first to see effects on tiny outposts

A Headline shortly after the hurricane.
From **The Glasgow Herald** *2 January 2003.*

The Location of the Disaster

The Solomon Islands lie east of Papua New Guinea in the western part of the Pacific Ocean, 1100 miles north-east of Australia (see Figure B). The country is small and economically insignificant. Even within the Solomons, the affected islands of Tikopia, Anuta and Fataka in Temotu Province are tiny, and their location, 900 km east of the capital, Honiara, makes them extremely remote.

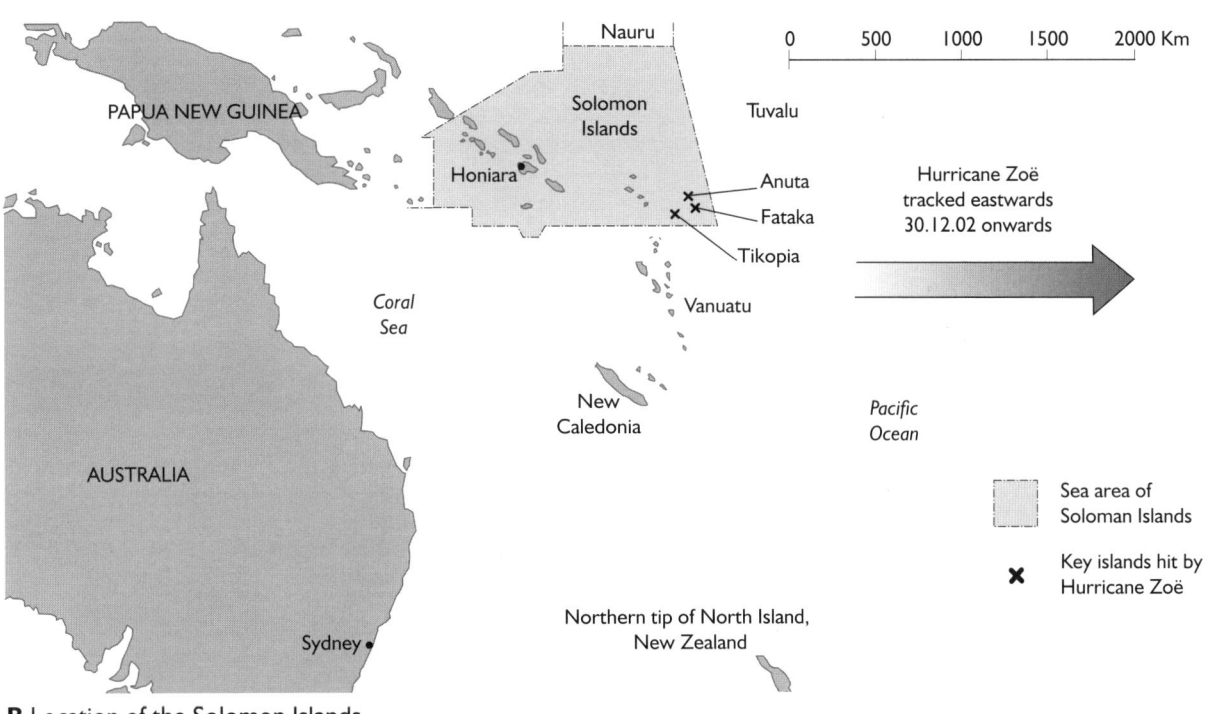

B Location of the Solomon Islands.

Assessing the Damage

Photographs printed in the British press on 2 January 2003 showed that these islands were almost totally swamped by the impact of Hurricane Zoë (see Figure C). The photos, taken by Geoff Mackley, a freelance photographer and filmmaker from New Zealand, showed walls and roofs ripped off buildings, trees shredded and toppled. The stumps of palm trees were all that remained of villages on Tikopia. The hurricane hit the tiny islands of Tikopia (the largest at just three miles long), Anuta and Fataka, and the 3000 islanders on Sunday, 29 December 2002. It was immediately classed as an extreme climatic event, with winds up to 225 mph and waves of around 10 m being recorded. Radio links to the outside world, the main form of communication, were immediately cut by the force of the storm. Geoff Mackley was the first outsider to visit the islands in the wake of the storm and his report states:

'Every tree in the island has been blown over or shredded, the island [Tikopia] is completely denuded of vegetation, almost every building has been damaged ……. The sea has come through some villages, burying them.'

Mackley had to send his report from Honiara, about 900 km away from Tikopia, on New Year's Day 2003 and it appeared in the media on the following day. He had been unable to land on any the 600 tiny islands in the Santa Cruz group of the eastern Solomon Islands as none of them has a landing strip, with or without the hurricane.

Just flying over the affected islands, as Mackley and subsequently the Australian Air Force did, was not sufficient to estimate the loss of life and the damage done to buildings, the community and its economy. Many islanders live in houses built of leaves and branches, although they had been given advance warning that Zoë was on its way. Reports made within the first few days of the disaster were therefore most unsure of the degree of damage experienced. The first flyovers saw islanders signalling and waving, with some apparently even going about their everyday business of farming and fishing. On the other hand, it was also obvious that considerable damage had been sustained. On Tikopia, two villages with a joint population of 700 people appeared to have been swept away. Dr. Hermann Oberli, the doctor on the first relief ship, certainly expected to find a significant number of people with injuries resulting from flying debris. In contrast to the estimates of significant damage on the islands, the hurricane weakened after hitting them and caused no further damage as it moved eastwards.

C Trouble in Paradise: The stumps of palm trees are all that remain of a village on the storm-ravaged remote island of Tikopia. **The Glasgow Herald** 2 January 2003.

When the relief ship arrived on 5 January, it was discovered that on Tikopia everyone had survived by sheltering in caves or in the mountains. The immediate crisis was the lack of drinking water, because this had been contaminated by seawater, and the lack of adequate sanitation, which could cause secondary disease. Electricity supplies were totally cut off and homes and crops were also largely destroyed.

An early estimate of the time needed for the economy to recover fully was put at two to three years. Quick growing crops, like potatoes and cassava, will regrow in two to three months, but most other cash and subsistence crops could take two to three years to become established and productive once again.

It seems that Tikopia will therefore be reliant on outside aid for at least that period of time. Anuta was not as badly damaged as had at first been thought. Ninety per cent of homes were barely damaged and 70% of subsistence crops survived. There were some injuries, but no one lost their life.

The Relief Operation

The government of the Solomon Islands is, at present, severely short of funds. £66 000 was made available for the relief operation. The Australian government also came up with early emergency supplies and extra funding, although they were criticised for not doing this sooner.

Exam Practice Questions

1 a) What conditions must occur in order for a hurricane to form? Page 216 in the pupil book will help you. (3)

b) Explain how a hurricane forms from the conditions you have described in answer to Question 1a. (4)

2 In which areas of the world do hurricanes occur? Think about latitude, name particular oceans and refer to Figure 13.12 on page 216 of the pupil book to help you. (4)

3 Referring to a hurricane you have studied, describe the types of damage likely to be caused by one. (6)

4 In what ways are the results of a hurricane in an LEDC different from those in an MEDC? (7)

Hurricane Zoë, 29 December 2002

These model answers are designed to show you how to get full marks for the longer questions in the Case Study Extra for this section. Read the question carefully and write your own answer. Then read the model answer and the examiner's notes to see what he or she is looking for when awarding the marks. Decide how many marks you think the examiner would give your answer. Decide how to change your answer to this particular question to increase your marks. What will you do differently next time you answer a similar question?

3 Referring to a hurricane you have studied, describe the types of damage likely to be caused by one.

(6)

> Hurricane Zoë occurred on 29 December 2002 and affected the eastern Solomon Islands of Tikopia, Fataka and Anuta. Winds of up to 225 mph were recorded, making it one of the strongest winds recently recorded. However, despite this, the damage caused was not immense.
>
> Tikopia was the island worst hit. Large numbers of trees were damaged, toppled over or shredded where they stood. Many palm trees are also crops, producing palm oil. The islanders are poor and most live by subsistence farming. Other crops were also damaged, meaning that aid will be needed for a year or more until the next season's crops can be planted and harvested. Some islanders live by fishing; boats were also lost in this hurricane. However, a typical boat is small and made from local materials, so is not too difficult or expensive to replace. Most people's homes are built of leaves and branches. They were also destroyed by the storm, but can be relatively easily rebuilt. Nevertheless, it is likely to take two to three years of aid to restore the economy fully.
>
> There were no deaths as a result of this hurricane, although a number of people were injured. The islanders had been warned of the hurricane's approach and because they are used to such events, they therefore took precautions. Most of them sheltered in caves or in the mountains. However, there was a risk of disease as drinking water supplies had been contaminated with seawater. Electricity was also cut off and was slow to be restored.

> Hurricane Zoë was chosen for this model answer, but you could use any other hurricane you have studied. The question asks for a description so there is no need for any explanation, which would not improve your marks and it would waste precious time in an exam. The facts have been organised logically and similar points grouped together; for example, deaths and disease are together in one paragraph and everything connected with the economy is in another paragraph.
>
> **Level 1 (1–2 marks):** *If you included a few facts, not very logically put together, you would achieve this level.*
>
> **Level 2 (3–4 marks):** *Here, you would have included more points, but without organising them.*
>
> **Level 3 (5–6 marks):** *All the points from the case study are included and they are grouped sensibly into paragraphs.*

4 In what ways are the results of a hurricane in an LEDC different from those in an MEDC? *(7)*

An example of a hurricane affecting an LEDC is Hurricane Zoë, which hit the remote Temotu Province of the Solomon Islands on 29 December 2002. Most people on these islands live a village existence based on subsistence agriculture. Primary damage included widespread destruction of homes and crops so people had to begin again by planting new crops, some of which take 2–3 years to become properly established. Outside aid will therefore be needed for a considerable period of time. These are poor communities without insurance. Help came to the Solomons mainly from Australia, the nearest MEDC.

In this event no one died as people were warned in advance by radio and most were able to shelter in caves and hills. Other LEDC hurricanes have brought higher death rates in more densely populated areas and caused secondary disasters such as severe landslides as in the case of Hurricane Mitch in Central America in October to November 1998. In this case, whole villages, their crops and livestock were swept away in the rainstorms that came with the hurricane. Floods and landslides killed thousands.

A contrasting example of a hurricane in an MEDC is Hurricane Andrew in the USA. Florida and North and South Carolina were affected. Houses were flattened by both winds and huge waves moving onshore. People fled inland, usually by car, taking their valuables with them. Public services, such as electricity, were returned more quickly than after either Zoë or Mitch and insurance policies allowed rebuilding to take place within a few months. However, a proportion of the population were without insurance and they were affected in a similar way to people in LEDCs. In MEDCs the time taken for rescues and repairs is much shorter than in LEDCs. Also emergency services are available and react much more quickly.

Improve Your Mark!

This answer uses Hurricane Zoë as an example, but also refers to Mitch, which was much more devastating even though its winds were less severe. Figures 13.15 and 13.16 on page 217 of the pupil book consider Hurricanes Andrew and Mitch and compare their impacts as examples from an LEDC and an MEDC. You could use any example involving Florida or another part of the USA.

Level 1 (1–2 marks): *A few points on each of two hurricanes would be made and the choice of events might be poor, e.g. you would be awarded only one mark if you mentioned only an LEDC or only an MEDC.*

Level 2 (3–5 marks): *Better choices would be made with some good points about each; however, a good comparison of the two sets of consequences would be missing.*

Level 3 (6–7 marks): *The quality of the comparison would be much better with lots of points and direct contrast between what is going on in the LEDC and in the MEDC.*

CHAPTER 14 deals with ecosystems, which is often a popular topic with pupils, and within it there are also some firm favourites, such as rainforests in particular. The pupil book places some emphasis on the rainforest as an ecosystem and the *Teacher's Resource Book* devotes two activity sheets either wholly or in part to this subject. The pupil book also includes three sections that discuss a wide range of ecosystems and biomes – deciduous and coniferous woodland, tropical grassland and Mediterranean zones. In addition, the case study concerns sustainability of forest in Malaysia.

A broad range of skills is needed for work with the materials in the *Teacher's Resource Book*, including comparison of two biomes, interpretation and annotation of graphs and photographs, completion of tables and, in particular, research from an atlas because the focus is on world-scale biomes, their locations and the explanations behind their distribution.

Ecosystems 230 – 231

The chapter begins with theoretical aspects such as the definition of an ecosystem and its basic characteristics and flows. Of the main GCSE and Standard Grade topics, British deciduous woodland is the ecosystem most likely to be familiar to pupils. Here they will study it from a systems approach.

Activity Sheet 14.1 looks at the functioning of deciduous woodland and employs a photograph to encourage pupils to investigate in more detail what might have seemed so familiar . It places the theory of nutrient cycling in the context of this particular ecosystem. Sustainability could be discussed as an issue in class.

World Biomes 232 – 235

Tropical rainforest, tropical savanna grassland, Mediterranean vegetation and coniferous forest are considered on these pages. Pupils will find a good summary of rainforests on page 232, which is then developed much further in the case study on pages 238–240.

Deforestation – the Destruction of an Ecosystem 236 – 237

The causes and consequences of rainforest destruction are tackled here in the context of the Brazilian experience.

Activity Sheet 14.2 explores the characteristic features of rainforests and their nutrient flows. This could be used as a basis for the discussion on rainforest destruction and sustainability suggested below as it is the breaking of the nutrient cycle that leads to long-term damage.

Activity Sheet 14.3 uses atlas work to locate both tropical rainforest and coniferous forest biomes. Pupils must describe patterns from a map, which is a useful skill. Question 2, which applies to coniferous forest, could be rewritten and used for rainforests as well. Threats to this ecosystem are considered.

Activity Sheet 14.4 considers characteristics of savanna vegetation. Links between climate, and vegetation and plant adaptations are emphasised.

Activity Sheet 14.5 takes a somewhat similar approach, but this time to the Mediterranean biome. A world map is provided on page 11 of the *Teacher's Resource Book* for location and the seasonality of the climate and its impact on vegetation are explored.

Case Study 14 Sustainable Forestry in Malaysia 238 – 240

Location, appearance, deforestation, methods of logging and moves towards sustainability are considered. Plenty of stimuli are provided for discussion and debate on these issues. Links with other parts of the syllabus can be made, for example the part played by tourism and the potential impact of clearance of rainforests on global warming.

Case Study Extra Sand Dunes at Shell Bay, Dorset

This case study presents a UK coastal sand dune ecosystem and so provides a link between this and other GCSE and Standard Grade topics, such as tourism and coastal defences. Several issues of exploitation and sustainability are tackled. Moreover this is an excellent topic for coursework investigations. The characteristics described here could equally well be researched on other coastlines.

1 Define the terms (i) *producer*, (ii) *leaf litter*, (iii) *biomass* and (iv) *nutrients*. (4)

2 From where does any ecosystem obtain its energy? (1)

3 Figure A shows an area of deciduous woodland in East Sussex.

a) What does the word *deciduous* mean? (1)

b) Match the phrases below to the photograph by putting the correct number
   in each box. (4)

   (i)   Dappled shade, allowing growth of ground cover plants.

   (ii) Path eroded and kept clear of vegetation by people walking along it.

   (iii) Leaf litter.

   (iv) Fallen tree, rotting as part of the nutrient cycle.

A Deciduous woodland in East Sussex.

4 Draw a diagram to show the nutrient cycle in a British deciduous woodland.
  Add the arrows going in the correct directions so that the cycle is completed. (4)

5 Read this article on woodland management in the UK.

> **UK woodland management condemned: Britain 'failing its forests'**
>
> Britain is fourth from bottom for forestry management among European countries according to ......
> WWF.* The UK has virtually no natural forests left and some ecosystem types have entirely disappeared
> .... Despite widespread loss, only 2.5% of the remaining area is under strict protection.

* WWF is the World Wide Fund for Nature, an influential global environment conservation charity.

a) List three reasons for the reduction in the total area of forestry in the UK at present. (3)

b) In what ways can Britain's woodlands be managed more sustainably in the future?
   (Hint: comment on the activities of the Forestry Commission and of environmental
   charities such as the Woodland Trust, government incentives to farmers to plant new areas
   of woodland and management of woodlands for recreation.) (6)

1 Figure A shows a typical tree in a tropical rainforest. The numbered labels refer to features that allow it to adapt to its environment.

   a) In which layers of the rainforest might you find such a tree? *(2)*

   b) Complete the table below to show:

      (i) which label belongs to which part of the diagram *(4)*

      (ii) the purpose of the adaptation. *(4)*

One line has been completed already for you.

Box number	Adaptation	Purpose of adaptation
	Tall straight trunk	
	Drip tips on leaves	
	Buttress roots	
1	Wide canopy	To have as much of the plant as possible reaching the light
	Branches only spread out high up the trunk	

2 Figure B shows how nutrients move in the vegetation–litter–soil cycle of the tropical rainforest ecosystem.

   a) Why are most of the nutrients in the ecosystem contained in the biomass (vegetation and wildlife)? *(2)*

   b) Why are so few nutrients stored in the soil? *(2)*

   c) Study Figure 14.10 on page 232 of the pupil book. Explain what happens to the nutrient cycle after people clear an area of rainforest. *(6)*

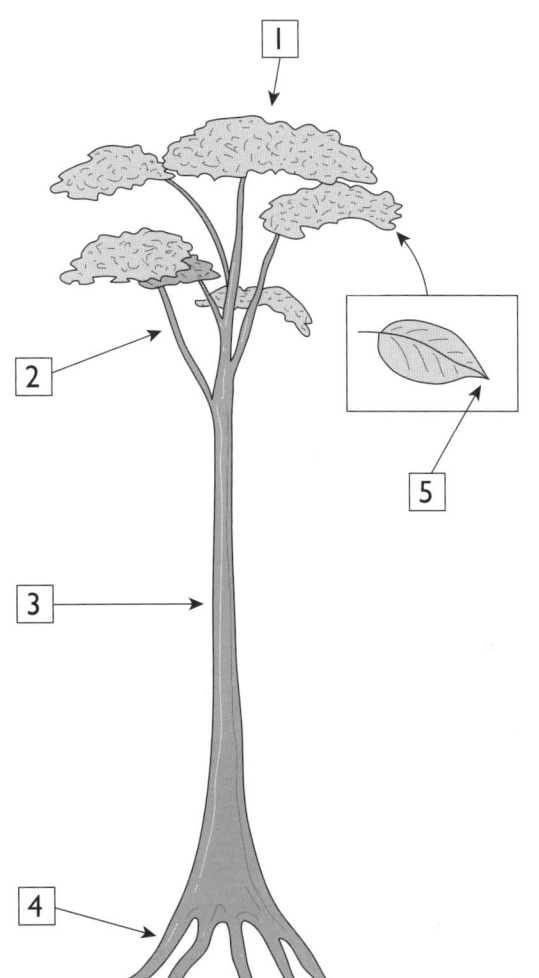

A The rainforest tree.

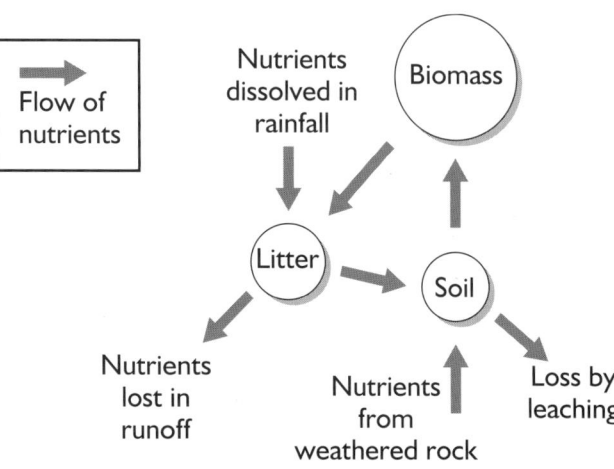

B The rainforest nutrient cycle.

1 Figure A shows the locations of tropical rainforest and coniferous forest.

 a) Using your atlas, name the lines of latitude marked on the map as numbers 1–5. Also note down the number of degrees and whether each line of latitude is north or south. *(10)*

 b) Describe the location of the rainforests. Refer to the latitudes and countries concerned. *(5)*

 c) How are the coniferous forest locations different from those of the rainforest? *(3)*

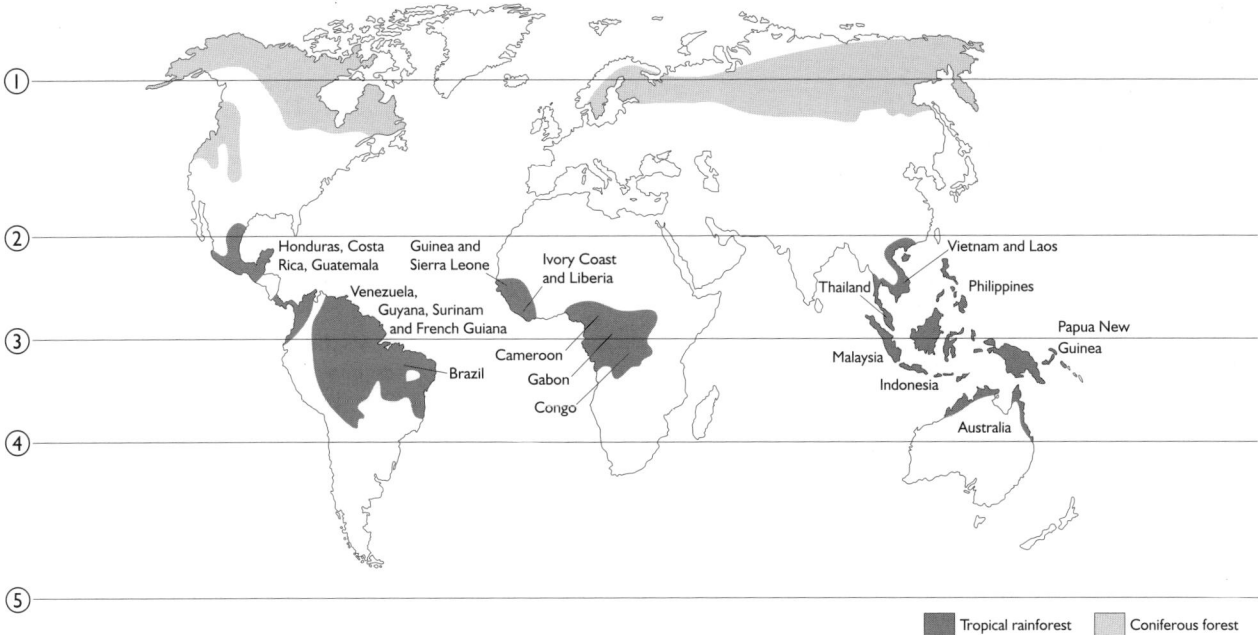

A Areas of tropical rainforest and coniferous forest around the world.

2 a) Name three uses of coniferous trees. *(3)*

 b) Do MEDCs or LEDCs usually produce and export coniferous timber? *(1)*

 c) Which type of country imports more of these products – MEDCs or LEDCs? *(1)*

3 Tropical rainforest is a threatened ecosystem because people choose to cut it down for several purposes.

 a) Place a tick beside any of the reasons in the list below that you think are a cause of major rainforest clearance. *(6)*

 - Production of energy, especially HEP ☐
 - Plantations ☐
 - Tourist facilities ☐
 - Production of hardwoods for export ☐
 - Shifting cultivation ☐
 - Extraction of minerals ☐
 - Wood pulp for paper mills ☐
 - Cattle ranching ☐

 b) Suggest three other activities that may result in clearance of the rainforest. *(3)*

 c) Which type of country (i) exports and (ii) imports tropical timber products? *(2)*

4 MEDCs could help in the conservation of tropical rainforests by doing the following:

 - cutting down on paper use
 - ecotourism
 - only buying furniture from sustainable forest products
 - debt swapping
 - rainforest action campaigns.

 In groups, choose three of these options and discuss how MEDCs could help the conservation of the rainforest. Be prepared to present your findings to the class.

1 Define the following terms, looking the words up to find a more detailed answer if need be.

　a) *xerophytic* (2)

　b) *seasonal climate*. (2)

2 Figure A shows the climate figures for an area of tropical savanna in East Africa. Describe the patterns of:

　a) rainfall (total, dry and wet seasons) (3)

　b) temperature (highest, lowest, range) (3)

　Quote figures from the graph to improve your answers.

3 Figure B shows a baobab (or 'bottle') tree, a thorn bush and some tufts of grass, all typical East African vegetation. Below, there are also some incomplete labels to show the adaptations of these plants to the savanna climate.

　a) Complete each label by adding the reason for that particular plant adaptation. (6)

　b) Draw in an arrow for each label on the diagram so that it points to the correct part of the appropriate plant. (6)

　c) Are savanna plants more adapted to the rainfall or temperature in their region? Give a reason for your answer. (2)

A Leaves are shed in the dry season to _____

B Tough thorny leaves to _____

C Very thick bark which can _____

D Wide shallow root system so rainfall can be _____

E Long tap root system so water can be _____

F Grass dies off in dry season because _____

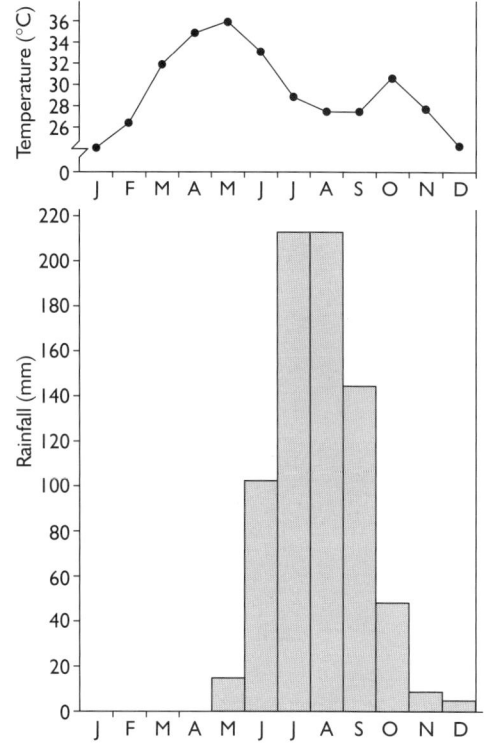

A Climate for a tropical savanna in East Africa.

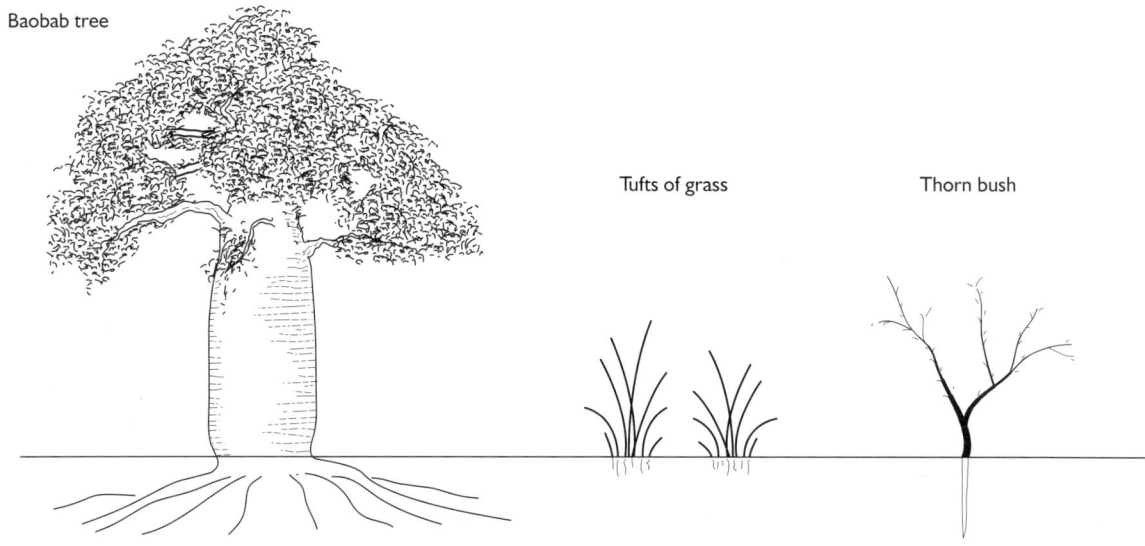

Baobab tree

Tufts of grass

Thorn bush

B Typical savanna vegetation.

1 Use your atlas to find the parts of the world that have a Mediterranean climate and natural vegetation.

   a) Mark these areas on the world map from your teacher. Make sure that you include a key.   *(5)*

   b) What do these areas have in common? Think in particular about latitude and position within the continent.   *(4)*

2 Figure A gives data for a Mediterranean climate in Sacramento, California.

Months	J	F	M	A	M	J	J	A	S	O	N	D
Temperature (°C)	7	10	12	15	17	20	24	23	20	17	12	8
Rainfall (mm)	95	70	70	40	20	5	0	0	4	15	40	95

A Mediterranean climate data for Sacramento, California.

   a) Use these figures to complete Figure B. The first two months have been done for you.   *(5)*

   b) Using the information on your completed graph and choosing terms from the panel, fill in the gaps in the sentence below:

   Mediterranean countries typically have _____ with _____

   followed by _____ .   *(3)*

   > • mild wet winters   • hot summers   • cool summers
   > • drought conditions   • dry winters.

   c) Explain briefly why areas with Mediterranean ecosystems around the world are so attractive to people.   *(4)*

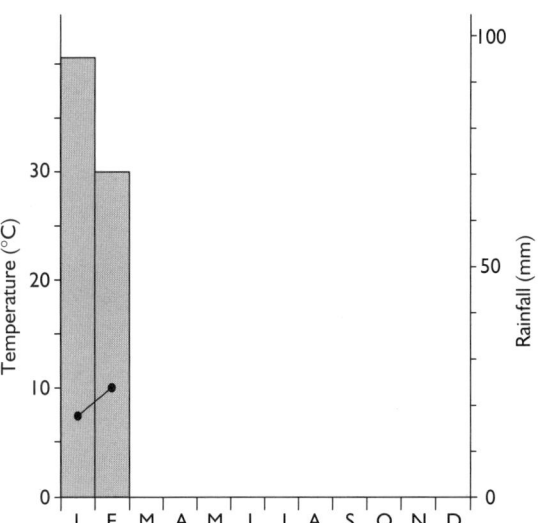

B Mediterranean climate graph for Sacramento, California.

3 Figure C shows an area of maquis vegetation on granite in Corsica.

  a) Describe the characteristics and density of the vegetation. (4)

  b) Give two differences between maquis and garrigue vegetation. (2)

  c) Discuss three typical economic activities that take place in the areas with
     a Mediterranean climate. (6)

C Maquis vegetation in Corsica.

Sand Dunes at Shell Bay, Dorset

Shell Bay is located just south of Poole Harbour in Dorset. The beach is backed by a broad stretch of sand dunes. The ecosystem here shows all the stages of dune development. A plant succession on a sand dune system is called a psammosere.

The land at Shell Bay is only a few hundred years old and Figure A shows its development. Once it was formed by longshore drift and spit development, vegetation took a hold, making the new land more stable. The direction of longshore drift is shown on the map. Today, the area is an important local and tourist beach and dune system, and the development of its ecosystem has been very much influenced by human activity.

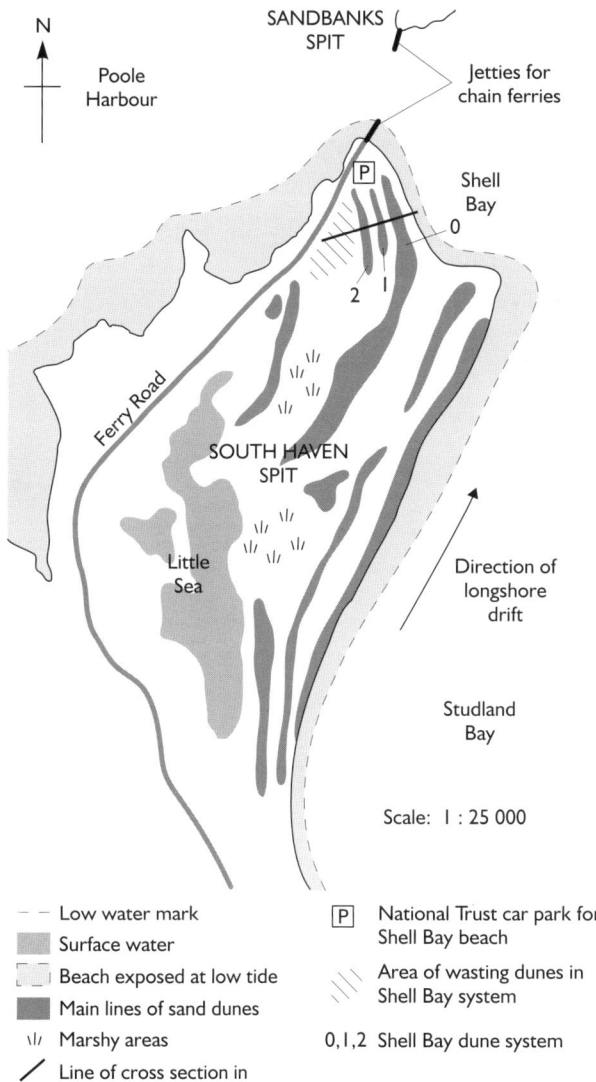

- - - Low water mark

░░ Surface water

▒▒ Beach exposed at low tide

■■ Main lines of sand dunes

\\/ Marshy areas

╱ Line of cross section in Figure B on page 164

P National Trust car park for Shell Bay beach

\\\ Area of wasting dunes in Shell Bay system

0,1,2 Shell Bay dune system

A The location of the sand dune system at Shell Bay, Dorset.

Dune Formation at Shell Bay

Sand dunes develop at Shell Bay because a certain set of circumstances exists:

- the wind blows onshore
- there is a long distance between high and low water marks, allowing a large stretch of sand to dry out and be picked up by the wind
- obstacles lie on the beach and sand collects around them.

The obstacles can be almost anything, either natural or man-made, such as vegetation, litter, a fence post, a litter bin, etc. Onshore winds pick up the dry surface sand and blow it inland. As it is deposited, the sand collects against these obstacles and starts dune formation.

The first dunes are called embryo dunes. There are always some of these on the beach at Shell Bay. They are so small that they can be flattened by someone simply stamping on them, but if they are left alone and a pioneering species of vegetation (the first ones to root in a new environment) takes root, they have a chance of growing into larger dunes. The most important species to grow on embryo dunes is sea lyme, which allows more sand to be trapped and to build up the dune further.

Further back from the shoreline, the heights of the dunes decrease and they become increasingly grey with humus. These older dunes are so far from the beach that no new sand is added to them, but the wind can blow some of their sand away. They are therefore known as wasting dunes as they are reducing in size. The lower hollow areas in between the lines of dunes are known as slacks. At Shell Bay these can be as low as the water table, allowing fresh water to sit on the surface and a greater variety of vegetation to grow. The further inland one goes, the closer the ecosystem gets to the climax situation. Succession is the development of the vegetation in an ecosystem, starting with the pioneer species to the largest and most sophisticated, i.e. the climax vegetation.

Marram Grass

Sea lyme quickly gives way to the most important sand dune species, marram grass, which can be found on the foredunes and first line of grey dunes at Shell Bay (see Figure B). It is similar in structure to sea lyme, but larger in scale. Marram grass has a huge root system, which grows down deeply to reach supplies of fresh water. The roots help to hold the sand together and so stabilise the dune. As the leaves trap more sand, the plant continues to grow and older stems become part of the system holding the dune together as it also grows. Marram grass is so successful because it is very well adapted to its environment. It has the following characteristics:

- it is halophytic, i.e. it tolerates a salty environment
- it is xerophytic, i.e. it tolerates an arid (extremely dry) environment
- its leaves curl in on themselves, like a tube, with the stomata inside, to reduce transpiration in this dry environment
- it has a very tough surface which is resistant to the wind.

Other Species of Plants

Different vegetation is found in the first slack. More shelter allows plants that are rather less tough to grow. Herbs, soft-stemmed green plants and rosette plants such as plantains are found. The rosette plants have no main stem; their leaves fan out from a central point, staying close to the ground. In this way they are protected from wind and can withstand being trampled by visitors' feet, a common problem along this stretch of coastline. Less tough grasses also grow in the slacks and several other species such as heather, gorse and brambles are common.

Across the wasting dunes, the climax vegetation is reached. Species are larger, but fewer in number, as some varieties are squeezed out. Tree species at Shell Bay include dwarf willow, alder, silver birch and various pines.

Changes in pH

The changes in the vegetation through the dunes at Shell Bay cause a change in the pH of the soil. The sand contains large quantities of broken shells making it, with a reading of up to 8.5, quite alkaline (see Figure B). The humus increases the acidity slightly, but the presence of acidic plants such as heather drops the pH reading as low as 4 in places, the most acidic a soil can be. Fieldwork readings taken every 10 m along the cross section show a steady reduction in pH readings over the dune system.

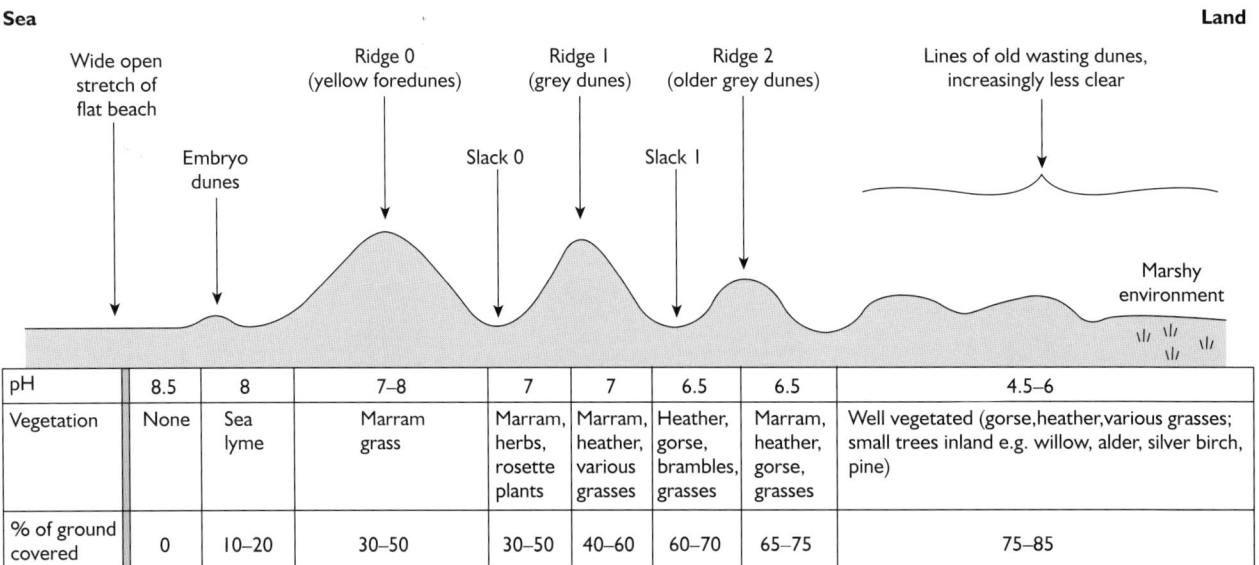

pH	8.5	8	7–8	7	7	6.5	6.5	4.5–6		
Vegetation	None	Sea lyme	Marram grass	Marram, herbs, rosette plants	Marram, heather, various grasses	Heather, gorse, brambles, grasses	Marram, heather, gorse, grasses	Well vegetated (gorse, heather, various grasses; small trees inland e.g. willow, alder, silver birch, pine)		
% of ground covered	0	10–20	30–50	30–50	40–60	60–70	65–75	75–85		

B Cross section through the sand dunes at Shell Bay, Dorset.

Human Activity on the Dunes at Shell Bay

The Dorset coast has a thriving tourist industry. Shell Bay Beach serves both tourists and the local population for such activities as walking, dog exercising, horse riding, kite flying, bathing, sunbathing and picnicking. Because the dunes themselves give some shelter from the wind people often sit within them and walk through them to find the best spots. Human feet are one of the greatest causes of dune erosion, particularly of embryo dunes. Burrowing animals, such as rabbits, also cause destruction at this location. Both people and animals have a destructive effect on the ecosystem.

Protection of the Dunes

Figure C shows one of the measures recently taken at Shell Bay to protect the embryo dunes from human feet. If the dunes are fenced off for a period of one to two years, they can develop their structure sufficiently to resist erosion better. Therefore, the main path from the car park to the beach now includes a wooden walkway, built partly to carry people over a very waterlogged section and partly to keep such a high concentration of feet off the dunes.

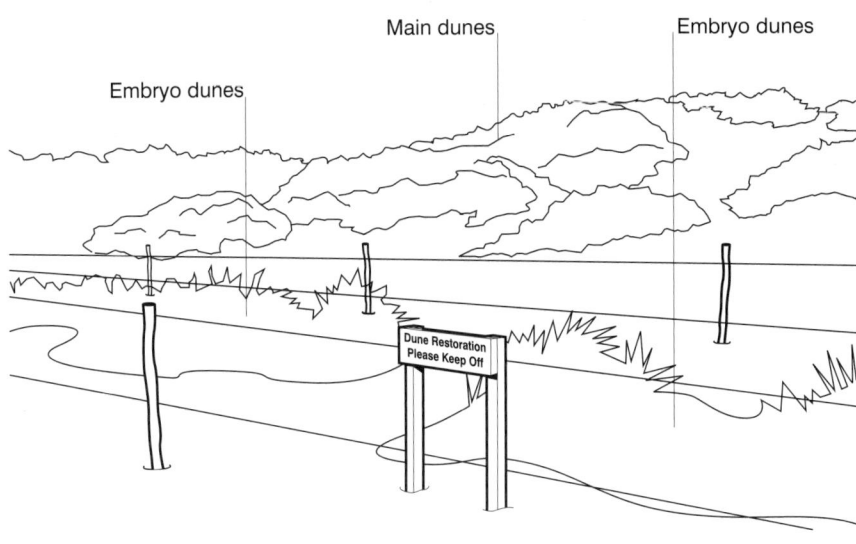

C Dune protection measures at Shell Bay, Dorset.

Exam Practice Questions

1 Using Figure B, describe the changes in vegetation found along the cross section through the dunes at Shell Bay. *(7)*

2 Explain why the pH of the sand/soil changes along the cross section. *(4)*

3 Discuss the advantages and disadvantages of tourist development in a sand dune area. You could discuss Shell Bay or another example you have studied. *(9)*

Sand Dunes at Shell Bay, Dorset

These model answers are designed to show you how to get full marks for the longer questions in the Case Study Extra for this section. Read the question carefully and write your own answer. Then read the model answer and the examiner's notes to see what he or she is looking for when awarding the marks. Decide how many marks you think the examiner would give your answer. Decide how to change your answer to this particular question to increase your marks. What will you do differently next time you answer a similar question?

3 Discuss the advantages and disadvantages of tourist development in a sand dune area. You could discuss Shell Bay or another example you have studied. *(9)*

> Tourism brings many advantages to the Shell Bay area. Primarily, money is brought in by tourists buying goods and services from local businesses. Demand increases, so new businesses may open, thus creating more employment, although some of these jobs are often seasonal in nature, which can be a problem.
>
> This area is close to both densely populated areas, such as Southampton, and large holiday resorts, such as Bournemouth. Therefore a huge demand exists for leisure sites. Shell Bay is a relatively natural site, where both locals and holidaymakers can enjoy an attractive landscape. The site is looked after by the National Trust, so that it is less likely to become spoiled.
>
> However, there are many disadvantages of such heavy human use. In really good weather, the beach can become too busy to enjoy. Animals can foul the beach, making it unpleasant. This is a popular area for walking dogs, as well as horse riding. The car park, though constructed of natural materials where possible (gravel floor, timber rails), is regarded by some as an eyesore. So many cars also lead to a concentration of air pollution, with the negative effect that has on the vegetation and animals.
>
> People walking through the dunes can have a number of negative effects. Wildlife is disturbed, a particular problem at Shell Bay because South Haven Spit is the only location in the UK where all six native reptile species can be found together. Human feet also erode the sand dunes, leading to the formation of blow out erosion, made worse when vegetation becomes trampled.
>
> One last point – litter, dropped by visitors, makes the dunes look untidy. On the other hand, litter, on occasions, can be an indirect advantage. It can trap sand, and in so doing, act as a starting point for the formation of a new dune in the system!

The exact detail in your answer will vary according to the example you have chosen, although many of the same basic points do apply to different areas. It is important not just to include points that could apply to any case study, such as the paragraph on litter. Also include factors that apply specifically to your chosen area. The answer above is about Shell Bay and some points, such as those about the wooden walkways and the reptiles in the ecosystem, are uniquely important in that location.

Whatever area is under discussion, a balance is required between the advantages and disadvantages suggested. Include at least three of each for a really good quality answer. Any extra points you want to make could be in either category. Explain the ways in which human pressure damages the dune ecosystem very clearly.

In a good discussion, the advantages and disadvantages can be discussed at the same time. The same issue could be viewed as positive in one sense and negative in another, so for example it is pleasant to walk dogs, but they may foul the beach.

Level 1 (1–3 marks): *An answer at this level consists of a list of any number of points without any discussion.*

Level 2 (4–6 marks): *Both advantages and disadvantages are included, although the balance may be one sided.*

Level 3 (7–9 marks): *A full discussion is needed, with a wide variety of points and a good balance between advantages and disadvantages.*

CHAPTER 15 includes eight sections covering the basic rock types, their formation and landscapes, the key elements of weathering and soil erosion and management. Two case studies on the Sahel and Pennine limestone landscapes both emphasise the effects of human activity.

Rocks – Types, Formation and Uses 📖 244 – 245

First, the origins of the three main rock types, igneous, sedimentary and metamorphic, are explained including a number of photographs to help illustrate the major differences between rock types. Characteristics of resistance and permeability are considered, and a table to summarise the uses of rocks introduces the concept of human involvement with the physical features.

Weathering and Mass Movement 📖 246 – 247

The principles of physical and chemical weathering are explored and a selection of weathering processes including freeze-thaw, exfoliation, weathering by tree roots and carbonation is then considered in greater depth. Carbonation links with the sections on limestone scenery on pages 248–249 and with the Case Study Extra. Page 247 considers the major categories of mass movement, illustrated by the disaster that befell Armero, Colombia in 1985.

Activity Sheet 15.1 complements this study by asking pupils to consider the crucial differences between weathering and erosion, an issue that often causes confusion. Questions emphasise freeze-thaw, exfoliation and biological weathering, the three types of physical weathering in the pupil book.

Limestone 📖 248 – 249 and 251

Figure 15.14 on page 248, with the photographs on page 249, link the formation of Carboniferous limestone landforms to their actual appearance, both on the surface and underground. This section is supported by the Case Study Extra on the Pennines.

Chalk and Granite 📖 250 – 251

There is a brief consideration of how the ways in which chalk and granite are formed have led to the creation of characteristic landscapes. Diagrams and photographs assist.

Activity Sheet 15.2 deals with chalk, revising the concept of permeability introduced on page 245. It considers the impact of a chalk landscape on human settlement using an example on the South Downs.

Activity Sheet 15.3 considers the formation of igneous rocks in general and granite in particular (see also pages 244 and 250 in the pupil book). The locations of granite areas and human activities are also included. The south-west of England is given as an example and use of atlases is encouraged as an extension activity for more able pupils.

The Mining and Quarrying of Rocks 📖 251

Human exploitation in the form of quarrying and mining is discussed in the context of its impact on national parks. The study on page 251 could be the start of a debate on the advantages and disadvantages of these economic activities.

Soils 📖 252 – 253

Soil formation, soil profiles and world soil types are considered here. Brown earths, podsols and tropical red earths are shown as detailed profile diagrams.

Activity Sheet 15.4 reinforces work on profiles in general and introduces ferruginous soils to tie in with the requirements of the main GCSE and Standard Grade specification. Comparison between profiles is encouraged using the two typical UK soil types, brown earths and podsols, shown in Figures 15.30 and 15.31 and the two tropical types, tropical red earths shown in Figure 15.32 on page 253 and ferruginous soils. Question 4c links to Chapter 14.

Soil Erosion 📖 254

Causes of soil erosion are considered. Figure 15.33 on page 254 summarises the factors concerned.

Soil Management 📖 255

Ideas for reducing soil erosion and salvaging damaged areas are explored here. Pupils living in, or visiting, a rural area could be asked to observe good and bad practice with regard to soil conservation.

Activity Sheet 15.5 compares soil management in Zimbabwe and the UK, an LEDC and an MEDC. Pupils are asked to make judgements about methods used in both.

Case Study 15 Desertification in the Sahel 📖 256 – 258

This study presents the causes and effects of desertification.

Case Study Extra Pennine Limestone Landscapes

This considers the physical limitations to human use of the Carboniferous limestone landscape in the Yorkshire Dales. Information on the soils, vegetation, hill farming, mining, quarrying and tourism are included. The number of human activities addressed is not exhaustive, but represents those occupying the most land and creating the majority of jobs in this part of the Pennines.

1 What is the main difference between weathering and erosion. (2)

2 Using the information on page 246 of the pupil book, fill in the missing terms in
the diagram below. (4)

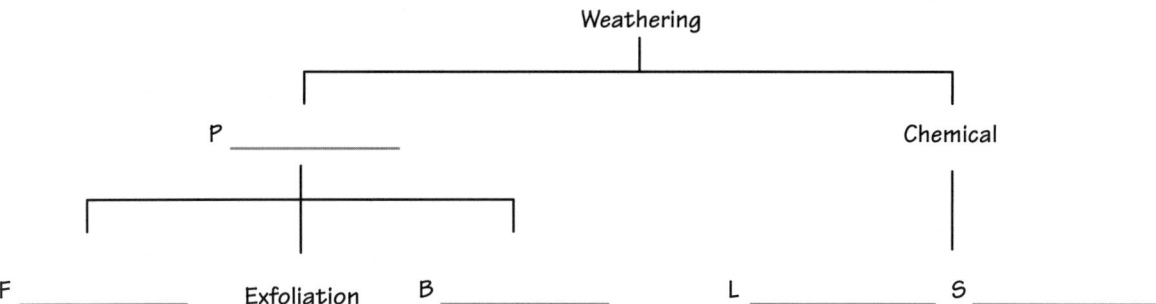

3 a) Figure A shows the process of freeze-thaw weathering (frost shattering). Complete the
labels by filling each of the four blank boxes with one of the five suggestions here: (4)

- Moisture trickles into a crack in the rock
- Crack is enlarged
- Snow falls into the cracks in the rock
- Rainfall or melted snow provide moisture
- Pieces of rock break off and fall down the slope

b) Name an area in which frost shattering is likely to happen frequently. (1)

c) Name two physical conditions that are needed for frost shattering to take place. (2)

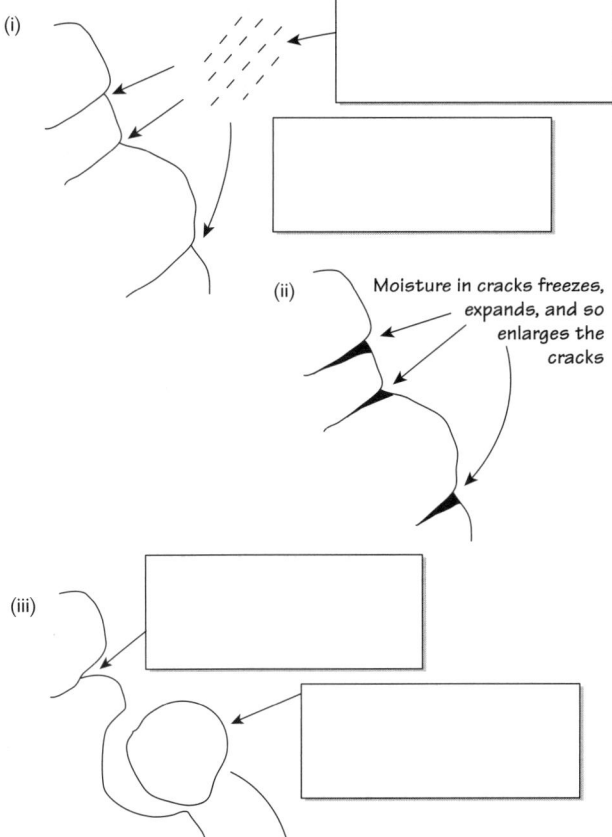

(i)

(ii) Moisture in cracks freezes, expands, and so enlarges the cracks

(iii)

A Freeze-thaw weathering.

4 Exfoliation is another type of weathering.

a) Give another name for exfoliation. (1)

b) In what type of region does it occur? (1)

c) Name such an area in the world. (1)

d) Describe what happens during
exfoliation. (4)

e) Draw a diagram or set of diagrams to
illustrate the processes. (6)

5 Biological weathering can be classed
as a type of weathering on its own or
it can be classed as part of physical
weathering. It results from the action
of plants, such as roots growing into
cracks in the rock and widening them,
or the action of animals, such as
burrowing by rabbits. Either draw a
diagram or give a written explanation
to explain the operation of both of
these types of biological weathering. (2 x 4)

1 Using page 245 in the pupil book to help you, define the following terms:

  a) *impermeable rock* (1)

  b) *permeable rock*. (1)

2 There are two types of permeable rock.

  a) Explain the differences between them. (3)

  b) To which of these types of permeable rock does chalk belong? (1)

3 Does chalk belong to the igneous, sedimentary or metamorphic group of rocks?
Give a reason for your answer. (2)

4 Figure A shows the area around Wilmington in East Sussex, which is on the edge of the
South Downs with a clay vale next to it.

  a) Wilmington village is built on a wet point site. Describe this type of site and explain how
it is linked to the local rock types. (5)

  b) Why did the original settlers of Wilmington choose a wet point site? (1)

  c) The annotations on the map include some possible factors that might have influenced
the original settlers' decision and also be economically useful today. Write one
paragraph to explain how this chalk landscape would have been useful to early settlers
and a second to discuss its uses today. (10)

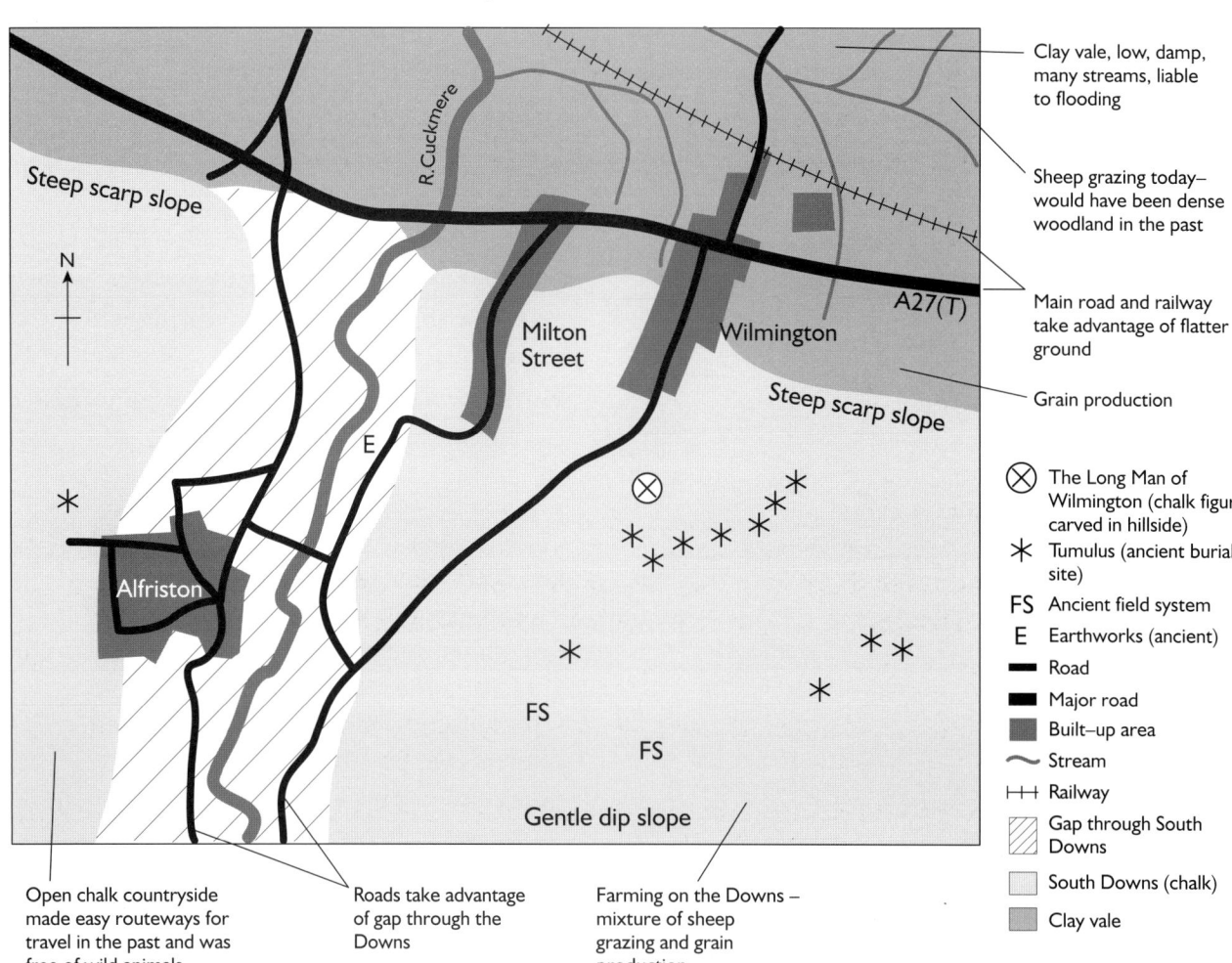

A The area around Wilmington, East Sussex.

1 Granite is an example of an intrusive igneous rock. Using the information given to you in the text and Figure 15.1 on page 244 and in Figure 15.7 on page 245 of the pupil book, complete the spaces left blank in the following passage. *(5)*

> An igneous rock is one which results from _____ _____. Granite is intrusive, which means it cooled _____ the Earth's crust. Cooling was therefore slow, so granite crystals are usually _____ in size.
>
> The other main type of igneous rock is extrusive. This means the magma broke right through the crust and so cooled on the surface. Its crystals are therefore small because cooling took place _____. An example of this type of igneous rock is _____.

2 Figure A shows some of the best known granite areas in the UK. They have been numbered 1–3 and their names are listed in the key. Use your atlas to match the number of each area to its name. *(3)*

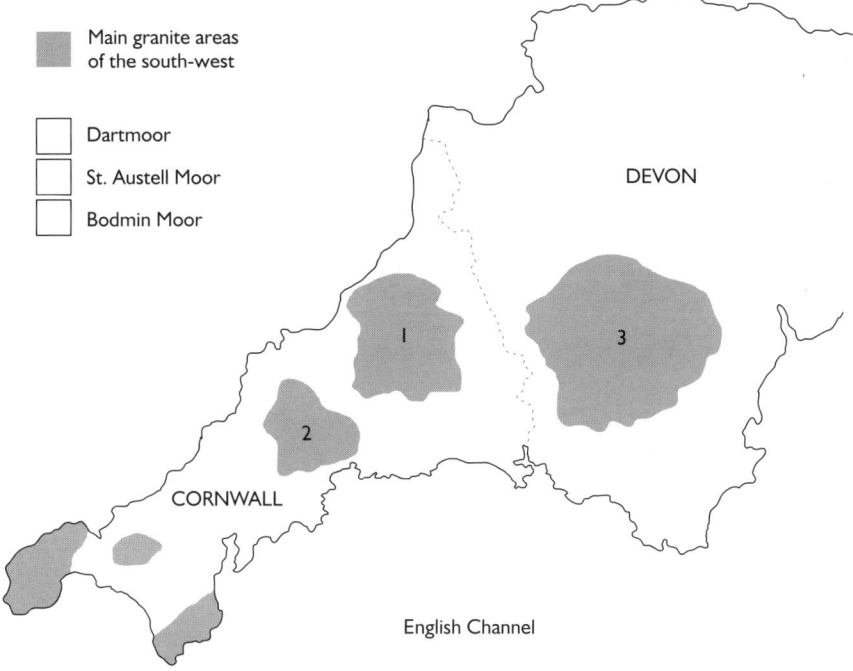

A Granite areas in south-west England.

3 South-west England is not the only granite area in the British Isles. Identify some other areas by looking up the British Isles geology map in your atlas. List at least three other granite zones. *(3)*

4 Granite landscapes are usually uplands with poor acidic soils and rough grass or moorland vegetation.

   a) Use Figure 15.7 to list five human activities that take place in such a landscape. *(5)*

   b) Farming is usually limited to sheep rearing. Explain why this is so. *(3)*

   c) Research ways in which the granite parts of the south-west are being used for tourism. Holiday brochures for the area might help you. Make a personal or group presentation to your class on this topic.

1 Explain briefly what is meant by the following terms:

   a) *soil profile*   (2)

   b) *horizon*   (2)

   c) *litter*   (2)

   d) *humus*   (2)

   e) *parent material*.   (2)

2 a) Five factors affect the formation of soil. Use Figure 15.27 on page 252 of the pupil book to help you complete the boxes in the flow diagram below, which show these factors.   (4)

   b) Explain the meaning of the term *Flora and fauna* in the diagram below. Figure 15.27 will help you here too.   (3)

   c) Add four ways in which climate is involved in soil formation to the diagram.   (4)

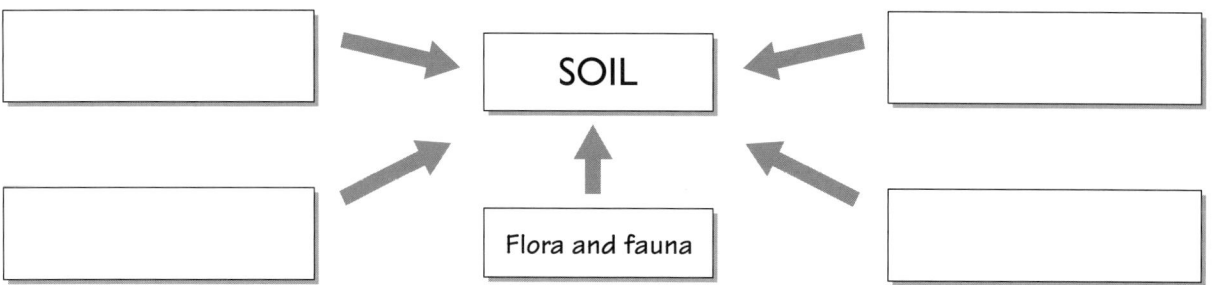

3 Study carefully the soil profiles of a brown earth and a podsol shown in Figures 15.30 and 15.31 on page 253 of the pupil book.

   a) Which of these profiles is deeper and by how much?   (2)

   b) Name one difference between the two leaf litter layers?   (1)

   c) What colours are the A horizons of the two soils?   (2)

   d) Give a detailed explanation of this colour difference.   (5)

   e) Which of the two soils has a hard pan? Explain how a hard pan forms.   (4)

4 Figure A shows a ferruginous soil found in the savanna areas in the tropical continental climate zones of Africa.

   a) Describe the horizons of this soil.   (4)

   b) Compare this profile with the one for the tropical red earth in Figure 15.32 on page 253 of the pupil book. Name one similarity and one difference between the two profiles.   (2)

   c) Tropical continental climates have wet and dry seasons. How do you think this will affect the soil profile? Page 213 in the pupil book will help you.   (3)

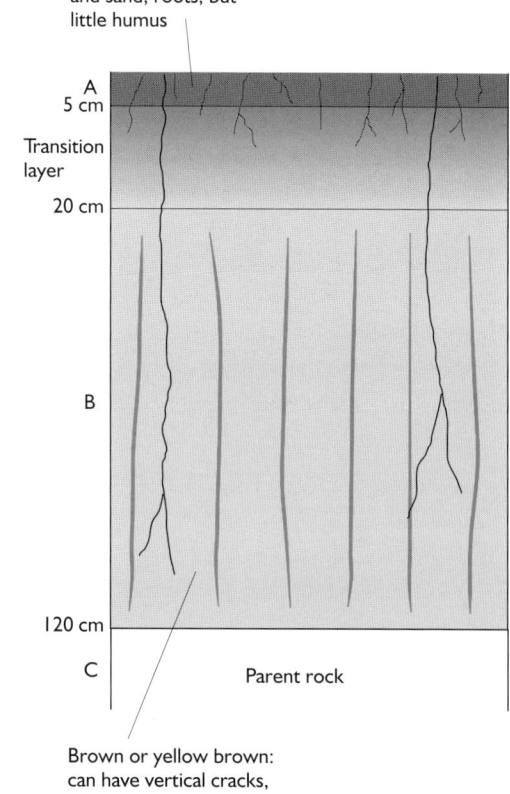

Very dark brown: clay and sand, roots, but little humus

A
5 cm

Transition layer

20 cm

B

120 cm

C   Parent rock

Brown or yellow brown: can have vertical cracks, some large roots

A A ferruginous soil profile.

1 What is meant by the term *soil erosion*? (3)

2 Read the article on soil erosion in Zimbabwe and then answer the following questions.

   a) State two physical reasons why Zimbabwe suffers from soil erosion. (2)

   b) Why have large-scale farmers had more success in preventing soil erosion than subsistence farmers? (2)

   c) Discuss the human factors that have made soil erosion worse in Zimbabwe in recent years. (6)

Soil erosion in Zimbabwe

An important management problem for any farmer is how to increase output over time without causing soil degradation. This is particularly hard in the tropics, where soils are more fragile and easily damaged.

Population growth puts extra pressure on the land because more people have to be supported by the same amount of land. Coupled with the intense convectional storms experienced in the tropics, which have the power to wash soil away, Zimbabwe's soils are really at risk. Using soils too intensively to produce food weakens the soils' structure, making them even more vulnerable to damage.

Much of Zimbabwe's farmland is used by village communities rather than owned by individual farmers. These communal areas have more severe soil erosion problems than land owned by individuals. In 1990, 27% of these areas were suffering from severe erosion and a further 32% had moderate erosion. Despite awareness of the problems, the situation became worse during the 1990s due to population pressure, the political situation in the country and the rapid spread of AIDS (which has a weakening effect on the working population). AIDS has affected up to 40% of young adults, male and female. Large efficient farms have been broken up into smaller units (as a consequence of the political situation) and handed over to subsistence farmers, who have less experience of coping with the needs of the land.

3 Figure A shows a number of ways in which UK farmers protect their soil. Fill in the boxes with suitable labels from the suggestions below. One has been done for you. (3)

- Shelter belt of trees
- Ploughing at right angles to the contours
- Replanting of hedgerows
- Ploughing parallel to the contours

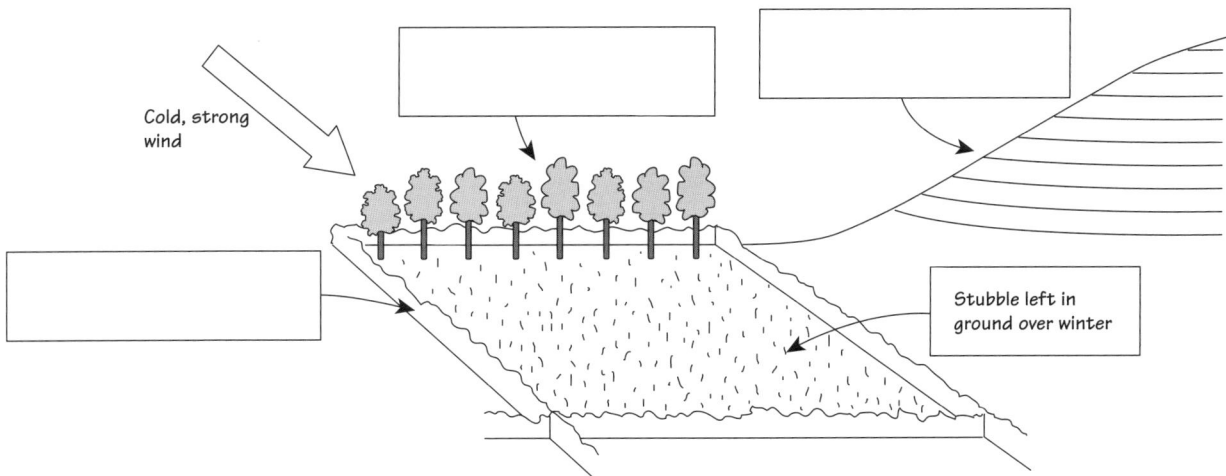

A Methods of soil management in the UK.

4 Which country has worse soil erosion – Zimbabwe or the UK? Give two reasons for your answer. (3)

5 Suggest ways in which farmers in Zimbabwe could reduce soil erosion. (4)

Pennine Limestone Landscapes

The Yorkshire Dales in the Pennines, the most significant Carboniferous limestone area in the UK, is well known for its open landscape. This wild landscape is the result of a combination of natural processes and human uses of the land.

A The location of the Yorkshire Dales.

B The Carboniferous limestone area in the Yorkshire Dales.

Soils and Vegetation

Limestone soils are dark brown rendzinas. They are very thin and mainly clay, which comes from the impurities left when the limestone is weathered away by carbonation (limestone solution). This type of soil also contains pieces of partly weathered limestone. Because it is such a thin soil, deep rooting plants, including trees, have little chance to thrive. In fact, woodland only really exists in any quantity when deeper areas of clay soils have collected in hollows. Rough grassland is more common and today is found over many areas of the higher hills. Waterlogging by lime-rich groundwater occurs in the valley bottoms and sometimes also on the lower valley sides, resulting in a fen (or marshy) environment.

Farming in the Pennines

The traditional hill farming landscape of the Yorkshire Dales is very distinctive. Hill farmers have special terms for the different types of land on their farms (see Figure 7.9 on page 100 of the pupil book for a particularly useful diagram):

- inland (or inbye land) consists of the valley floor
- intake (or allotment) is the lower valley sides
- open or high fell, the higher unimproved areas.

The first two areas, which have often been improved, are divided up and belong to individual farms. However, the open fell land is held in common (i.e. sheep wander across it as they wish, their ownership identified only by coloured markings on their coats).

Special features of this farming landscape include an intricate network of dry limestone walls, stone being the most easily available building material because much lies loose on the ground. As Carboniferous limestone weathers mostly along its joints, blocks break off and can then easily be used for such purposes. The lack of trees also means less timber is available for fencing and the thin soils discourage hedge construction.

Hill sheep farming maintains the open landscape. It is classified as marginal farming because it is not easy to make a good living from it. The physical difficulties limit what people can do with the land. Many farms only survive because:

- they have diversified into other activities such as tourism, for example offering bed and breakfast
- they receive UK government and EU subsidies.

Mining and Quarrying

In the past, the Yorkshire Dales were exploited for mining. Lead was mined from Roman times until the late 19th century, when veins became exhausted. Today the limestone itself is quarried. This is much larger in scale than the previous mining activities and has a much greater impact on the landscape. One quarry on the edge of the small town of Ingleton lies opposite White Scar Caves, an important tourist attraction. A relatively large concern, it causes certain problems and some solutions have been tried. Pressure exists to reduce the impact of quarrying on the landscape as much as possible.

Problems caused by quarrying

- A regular flow of heavy lorries.
- It makes the landscape unattractive for tourists.

Solutions tried

- The site itself has been carefully screened from view by a bank of trees.
- Permits are necessary for quarrying to take place.
- There are rules to limit the impact on the landscape.
- The company must reclaim the site after quarrying has ended.

Quarrying in the Pennines is unlikely to end in the foreseeable future because limestone is an important resource for the chemical industry and it provides jobs in a region which traditionally has high unemployment. Money earned from such business then flows into the whole local economy and supports other businesses.

Tourism

To date, tourism has been the most profitable human activity in this difficult Pennine limestone landscape. Attractions can be classified in four main ways: physical (geological and wildlife) and human (historical and crafts/local products). Examples are shown on Figure B on page 174 and their diversity is obvious. New ventures are common, such as the Grassington Festival of Classical and Jazz Concerts, which began in 1996.

The attractions of the Malham area

An increasing proportion of attractions have nothing to do with the physical environment or traditional way of life of the area. This could be criticised in one sense, but if such attractions take pressure off this fragile landscape and still enhance the local economy, then perhaps they are a good thing.

Some sites are certainly 'honey pots' and are being seriously damaged as a result. Malham village is one such example. The limestone scenery to the north of Malham village, with Malham Cove, Gordale Scar, several waterfalls, gorges, dry valleys and limestone pavements, is probably the most spectacular in the UK. Walkers, climbers and those who simply enjoy attractive open landscapes flock to this area (see Figure C). Those interested in history and archaeology also come because, for example, there is evidence of Iron Age and Mediaeval settlements and agriculture around Malham Cove.

Limestone crags

Scree

Thin soil and rough grass

C The foot of Gordale Scar.

Landscape damage at Malham

The capacity of this village and its surroundings to cope with the sheer numbers of people and their vehicles is limited, and is already being exceeded. The lane leading up to the village often consists of a long line of parked cars, rendering it single track and unattractive. The village shops no longer really serve the local population, but rather the tourists. Some residents have even turned their front rooms into tearooms in the summer.

The feet of walkers have caused severe wear and tear. The main path to the cove has been strengthened with limestone rubble, creating a white scar across the landscape. Some visitors remove chunks of rock, often from limestone pavements, to take home. This is causing noticeable damage to these features.

Carboniferous limestone produces a landscape that is difficult for people to exploit. Great care needs to be taken to use, but not to abuse, that landscape, if it is to be conserved for the future.

Exam Practice Questions

1 a) Describe the distribution of Carboniferous limestone within the Yorkshire Dales National Park. *(4)*

 b) To what extent do the limestone areas coincide with the major tourist sites? *(3)*

 c) Why do you think this is so? *(4)*

2 What limitations and opportunities do Carboniferous limestone rock and its soil place on people's use of the landscape? *(9)*

Pennine Limestone Landscapes

These model answers are designed to show you how to get full marks for the longer questions in the Case Study Extra for this section. Read the question carefully and write your own answer. Then read the model answer and the examiner's notes to see what he or she is looking for when awarding the marks. Decide how many marks you think the examiner would give your answer. Decide how to change your answer to this particular question to increase your marks. What will you do differently next time you answer a similar question?

2 What limitations and opportunities do Carboniferous limestone rock and its soil place on people's use of the landscape? *(9)*

> Carboniferous limestone has physical characteristics that limit its use. The fact that it is so pervious means that there are few surface streams. Most of the drainage goes underground. This limits agriculture. Much of the land surface is steep and some consists of bare rock. The relatively high alkalinity of limestone (pH 8–9) forms an alkaline soil which does not suit all vegetation. Also, the soil is thin and so does not lend itself to ploughing and crop rooting. Therefore sheep grazing is one of the few possible types of farming. Many farmers in this area only remain in business with the help of EU and UK government subsidies, which they receive because of the difficulties of farming in this landscape. Farmers are very important in conserving our present landscape. Without them it would become completely wild again.

> Steeper slopes and bare rock also limit settlement. Villages are usually found at wet point sites where the limestone lies next to impermeable rock. Although the landscape was more wooded in the past, the lack of woodland today leaves an exposed and windy landscape, which is less attractive for settlement than other lower surrounding areas.

> Perhaps the greatest economic opportunities lie in quarrying and in tourism. These take advantage of the rock type and its resulting scenery, and they contribute more to the local economy than other economic activities do. Walkers are attracted by the spectacular limestone features, open landscapes and views. Underground drainage systems, with caves, caverns, stalactites and so on provide other obvious tourist attractions. Tourist services such as hotels, bed and breakfasts, gift shops and tearooms exist to take advantage of the business generated. They create many jobs, although many of these have the disadvantage of being seasonal. Negative aspects of tourism include footpath erosion, such as at Malham Cove, traffic congestion and accidents, and overdevelopment at 'honey pot' sites.

> Quarrying provides less employment, but nevertheless means more money comes into the local economy. This means shops and services are supported, so creating other jobs and a higher standard of living for local people. However, quarrying has a significant negative impact on the landscape in terms of dust, noise, traffic and scarring of the landscape, as at the quarry opposite White Scar Caves near Ingleton. However, the presence of the quarry does not seem to discourage visitors to the caves, perhaps because they are more interested in what there is to see underground and because the site is well screened from the road.

Improve Your Mark!

This question demands one of the longest answers you will be asked to write at GCSE and Standard Grade level. It clearly asks for both limitations and opportunities to be considered. To achieve top marks, answers must have a balance between these and discuss the advantages and disadvantages of each type of economic development.

Level 1 (1–3 marks): *Mention of only one or two economic activities would achieve this mark; for example, you might have concentrated on opportunities or limitations, but failed to cover both.*

Level 2 (4–6 marks): *More economic activities would be included in this answer, but perhaps they are mostly opportunities or limitations without the appropriate balance between them or perhaps the advantages or the disadvantages of each activity have been emphasised, again without the balance required.*

Level 3 (7–9 marks): *A range of different types of economic activity is needed for a top mark answer. The model answer considers four activities, which, although not exhaustive, give sufficient range to produce a balanced and sufficiently long answer. Both the advantages and disadvantages of each activity are discussed.*

CHAPTER 16 is divided into six sections: tectonic activity, types of plate movement, causes and effects of volcanic eruptions in MEDCs and in LEDCs, and causes and effects of earthquakes in MEDCs and in LEDCs. Case studies in the chapter include Mount St. Helens in the USA, Merapi in Indonesia, Kobe in Japan and Takhar in Afghanistan.

Tectonic Activity 262 – 263

This section introduces the terms and concepts of plate tectonics, earthquakes and volcanoes.

Types of Plate Movement 264 – 265

This sections clearly illustrates the processes in operation at the four types of plate margin.

Activity Sheet 16.1 reinforces the concepts behind the theory of plate tectonics. Four outline block diagrams are used to check pupils' understanding of the types of plate margins. Pupils should be able to draw labelled sketch versions of the block diagrams.

Causes and Effects of a Volcanic Eruption in an MEDC 266 – 267

This section uses the famous example of Mount St. Helens in the Cascade mountain range of the north-west USA, which provides ample opportunity for atlas work. Pupils could be asked to locate Mounts Shasta, Hood and Rainier in addition to Mount St. Helens, and to calculate the length of the Cascade range from Mount Shasta to the Canadian border. There are some excellent resources on the Internet dealing in great detail with the Mount St. Helens eruption. Go to the website www.olywa.net/radu/valerie/mshafter and use the links available there to start your search. Alternatively any Internet search engine will provide a wealth of links.

Activity Sheet 16.2 builds on the study of Mount St. Helens in the pupil book by focusing on the process of renaturalisation of the eruption's blast zone.

Causes and Effects of a Volcanic Eruption in an LEDC 268 – 269

The volcano of Merapi, the most active in Java, Indonesia, provides the focus for this section. It is an excellent illustration of the popular perception of a natural hazard, since this volcano has killed people on ten separate occasions since 1920, and yet the local population always returns to live on its slopes. Further information can be obtained from these websites:
www.ipgp.jussieu.fr/~beaudu/merapi.html
www.vsi.dpe.go.id/mvo/generalview.html

Causes and Effects of an Earthquake in an MEDC 270 – 271

The Kobe earthquake of 1995 provides the focus for this study. Useful additional information can be obtained from the websites
www.city.kobe.jp/cityoffice/15/020/quake/index.html
www.zephryus.demon.co.uk/geography/resources/earth/kobe.html

Activity Sheet 16.3 centres on an eyewitness account taken from the website
http://ccs.cla.kobe-u.ac.jp/Asia/Visitor/Furm/
This is an especially compelling website. It features first-hand accounts from Japanese university students, compiled by Mike Furmanovsky, Professor of American Studies. The broken English adds immediacy to the story. Here is a taste of the experiences that you can find at this website from Toshiko Sasaki.

Suddenly, unexpectedly, and with no precaution, Kobe has fallen into ruin after the 20 seconds shake. I was in my apartment on the third floor of a four-storey-building, located in Kitano-cho, about 15 minutes walk up the hill towards north from the city center: Sannomiya. First, I felt the floor sank with a loud bang and then, the violent shake accompanied by terrifying sound followed: glasses in the cupboard fell on the low table with a clatter, something moved with a thud in a corner of my room. I covered up myself with my futon but still my body was bouncing out of control. The only thing I could do was to think desperately, 'How long on earth does this horrible shake last!?' It had lasted 20 seconds according to the news that I heard later, but I had felt as if it was 40 seconds or more! Electricity went off right after I had felt the first bang, so I was able to see nothing in my room in the dark. Right after the quake, I looked out the window to see how it was outside, the night view of Kobe was not lightened by the usual neon signs and electric lights but by fire!! Orangeish smoke was going up here and there, and those areas were lighting up the sky! Fire engines' sirens roared but they seemed to be lost. They were not reaching those places? Fortunately, my telephone line was alive so I tried to call my family but nobody answered. I wanted to see more things with my glasses but they were on the low table which I knew would be full of scattered broken glasses. I groped around me and tried to go to the next room because I remembered there should be my flash light on the shelf inside the front door and my headphone radio in my bag. I felt some glasses broken on my hand when I reached near the sliding door that separates the rooms....

Activity Sheet 16.4 includes statistics comparing the casualties from eruptions with those from earthquakes. Whilst there have been some eruptions causing extensive casualties, earthquakes are more deadly, largely because there is no warning. The pupils mark locations of eruptions and earthquakes on copies of the world map on page 10 of this *Teacher's Resource Book*.

Causes and Effects of an Earthquake in an LEDC
📖 272 and 273

This section includes a diary of the effects of the Afghanistan earthquake of 1998. This can form a valuable model for pupils to use for any major earthquakes or volcanic eruptions occurring during their course. Newspapers, TV and radio reports and Internet sites can provide information to build up a diary of the events and effects.

Activity Sheet 16.5 focuses on the Bhuj earthquake in Gujarat, India in 2001 and includes another eye-witness account.

Case Study 16 Volcanic Eruptions and Earthquakes in MEDCs and LEDCs 📖 274

This case study involves a summary comparison of the prediction of and preparation for earthquakes and their varied effects in MEDCs and LEDCs.

Case Study Extra The Eruption of Mount Pinatubo

This eruption in the Philippines in 1991 provides an excellent example of a successful human response to the threat posed by a natural hazard. As an LEDC, the Philippines might have been expected to face more difficulties in preventing casualties, but the country has a highly developed volcanological research institute (the Philippine Institute of Volcanology and Seismology) and was able to call on help from US volcanologists from the US Geological Survey. The USA had air and naval bases on the Philippines and was anxious to protect them.

1 a) What is a tectonic plate? (2)

b) What causes the plates to move across the surface of the mantle? (1)

c) How fast do the plates travel across the earth's surface? (1)

d) What events occur at plate margins? (3)

e) (i) Name the two types of crust of which plates may consist. (2)

(ii) How do the two types of crust vary and how does this explain the variation in tectonic processes at plate margins? (4)

2 The outline block diagrams in Figure A show four types of plate margin. Write each one of the following labels next to the correct letter on the diagrams. (4)

- Island arc of volcanoes
- Spreading ridge
- Subduction zone
- Crust
- Fold mountain range
- Ocean trench
- Mantle
- Transform fault

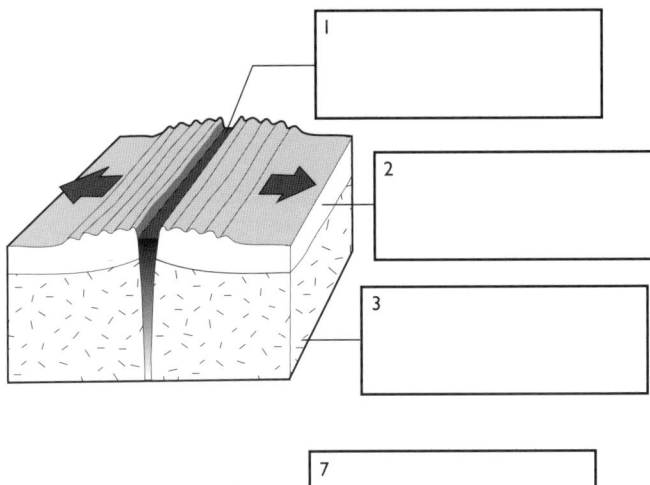
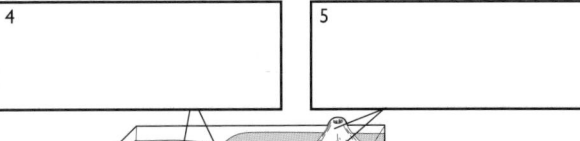
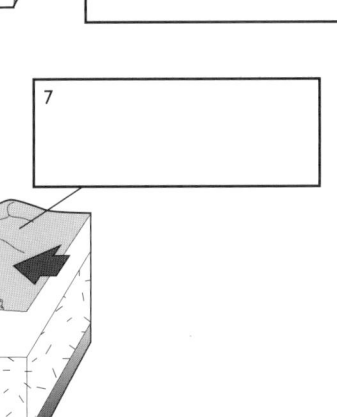
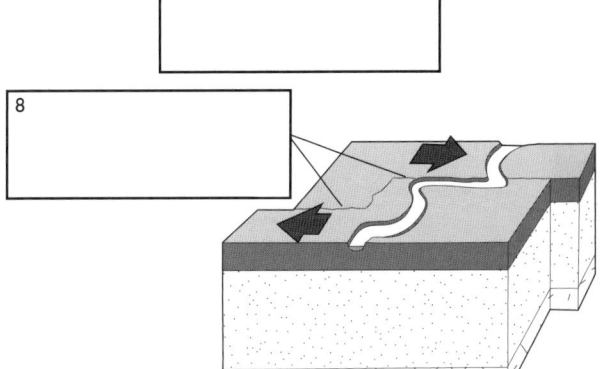

A Four types of plate margin.

3 a) What type of plate margin is found at the Mid-Atlantic Ridge? (1)

b) How is the Mid-Atlantic Ridge formed? (2)

4 a) At a destructive plate margin, what determines whether a subduction zone or collision zone is formed? (2)

b) Name an example of (i) a subduction zone and (ii) a collision zone. (2)

5 a) Explain the term conservative margin. Name an example. (2)

b) Why is there no volcanic activity, but much earthquake activity, at conservative margins? (4)

Read the description of the eruption of Mount St. Helens in north-west USA in 1980
on pages 266–267 in the pupil book.

1 a) Name the type of plate boundary shown in Figure A.   (1)

   b) Identify each of the numbered features on the diagram with the correct term in this list.   (8)

- Earthquake focus   ☐
- Volcano   ☐
- Rising magma   ☐
- North American plate   ☐
- Juan de Fuca plate   ☐
- Oceanic crust   ☐
- Continental crust   ☐
- Magma reservoir   ☐
- Mantle   ☐

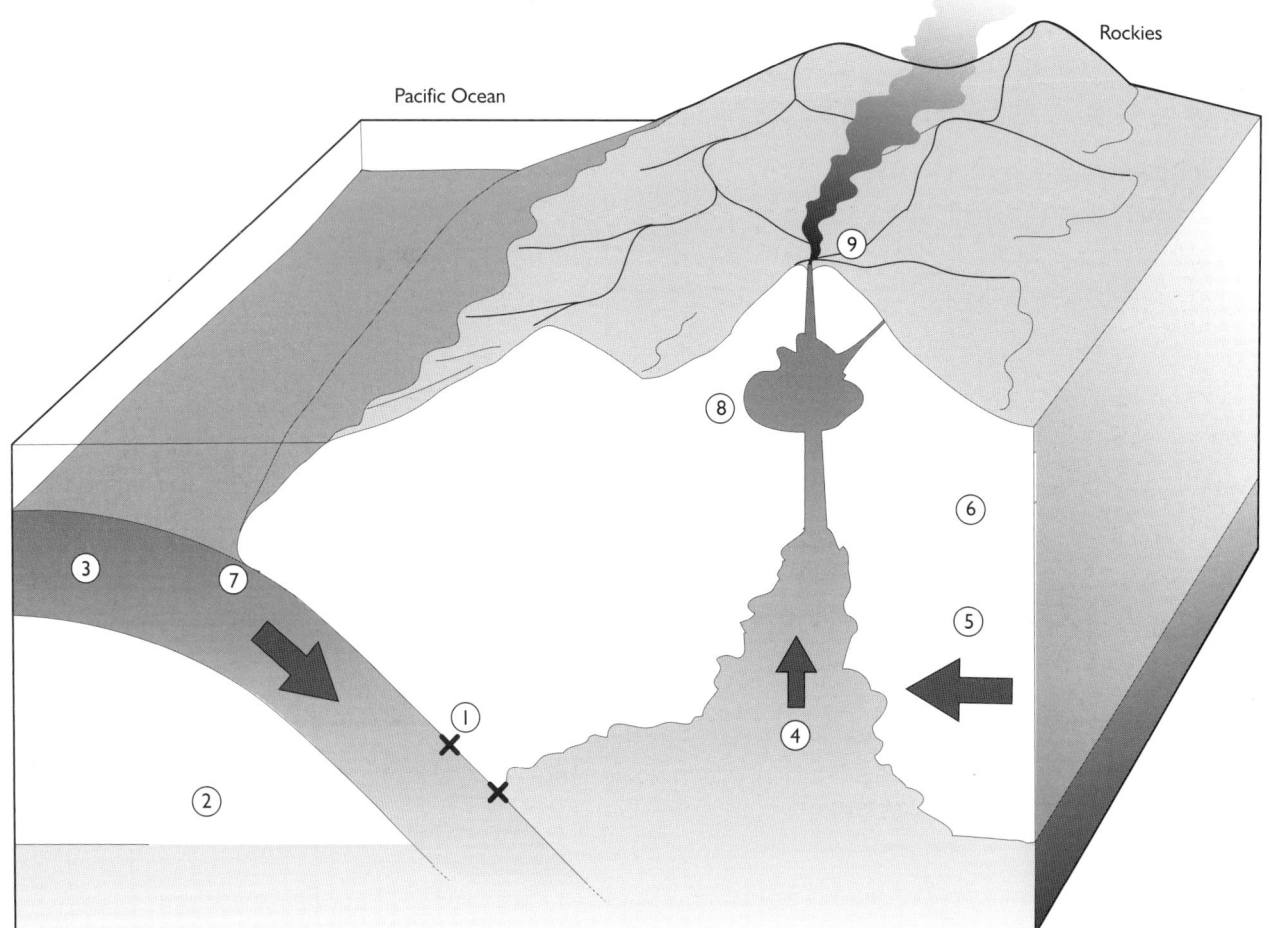

A The tectonic plate boundary responsible for the eruption of Mount St. Helens in 1980.

2 a) How does continental crust differ from oceanic crust?   (2)

   b) Why are these differences important?   (3)

The eruption of Mount St. Helens created a barren wasteland around the mountain. However, as the years passed, nature began to reclaim this wilderness. Read the article and the newspaper extract below.

On 26 August 1982, the US Congress established the area around the mountain as the Mount St. Helens National Volcanic Monument. Today this is one of the north-west USA's top tourist attractions. An area of about 110 000 acres has been preserved so it can heal without any human interference. While there have been several Visitor Centres built along the new highway into the blast zone, guests are expected to leave the land as they found it. The monument has become a living laboratory in which scientists and the public can observe the amazing return of life to the area.

Scientists say it will take up to a century for the blast zone to recover fully. However, over two decades later, the land blooms once again. Fish are swimming in the lakes and rivers. Wildflowers can be found bending in the mountain breezes. Elk and deer can be found in the hills, in greater numbers than before the eruption. Wildflowers are flourishing. Wildlife now living on the fringe of the blast area, from insects to birds, coyote, and elk, today move through the blast zone. Their waste products provide seeds, which germinate in the fertile volcanic debris. Thus islands of vegetation appear, which provide food and cover.

Just as flora and fauna are returning to Mount St. Helens, so are human visitors. The Spirit Lake Memorial Highway is home to five major visitor centres, a sprinkling of souvenir shops selling glass ornaments and pottery made from local ash, and a few eateries dishing up 'lava dogs' and 'volcano burgers'. Others head for the mountain's barely altered south face, where about 13 000 adventurers a year make a gruelling half-scramble, half-climb to the rim.

But some who earn a living from Mount St. Helens say that while current curiosity seekers may be more numerous than those who came before the mountain exploded, they're not as profitable. Faced with a cloudy day (predominant in this part of south-west Washington), many would-be tourists simply keep driving up Interstate 5 – leaving business owners pining for the days when weekenders from Portland thronged to cabins and campgrounds along the shoreline of now-devastated Spirit Lake.

'After 20 years, the novelty has worn off a bit,' says restaurant manager Don Shaw, who says revenue at his Hoffstadt Bluffs Visitor Center is down by 50% over the past few years. 'But if the mountain started erupting again tomorrow,' Shaw says, 'you'd see a cavalcade of cars up here. Let's face it: We all like disasters and dirty linen.'

From **USA Today** *11 May 2002.*

3 a) What is the Mount St. Helens National Volcanic Monument? *(1)*

b) Why has it been established? *(3)*

4 a) What evidence is there that the blast zone around St. Helens is recovering? *(3)*

b) How long is it likely to take for the blast zone to recover fully? *(1)*

c) How has the eruption changed the pattern of tourism in the area? *(3)*

Read the case study on pages 270 and 271 of the pupil book.

1 a) When did the Kobe earthquake occur?   *(1)*

   b) What was the cause of the earthquake?   *(2)*

   c) What were the effects caused by the earthquake?   *(1)*

   d) What other primary effects resulted from the earthquake?   *(4)*

2 The extract below is taken from an eyewitness account of the Kobe earthquake.
   It is written by Toshi Matsui, a Japanese university student. Imagine that you
   were a newspaper reporter in Japan during the Kobe earthquake. Write an article
   for your newspaper, describing the events and providing a simple explanation of
   the cause of the earthquake. You should include diagrams in your explanation.   *(10)*

3 What measures were taken after the Kobe earthquake to reduce the effects of
   a future earthquake?   *(6)*

On the night of the earthquake I was lucky not to be alone because my club mate had stayed in my home. When the Hansin earthquake shook Kobe, we slept. As soon as my friend got up, she shouted 'Hey! There is an earthquake!' We went out of my room slowly wearing our pyjamas. A young woman cried 'Help! My grandfather and parents are still in my house! My house is broken.' Men ran up and they saved the people from being buried alive, but we just trembled. We were too afraid to say anything. I wished I was dreaming, but this incident was not a dream, but reality.

The sun rose and a sad sight stunned us. Two storey houses became one storey houses, the telegraph and electricity poles had been snapped and the ground was full of cracks. The fire flared up and many fire engines ran around Kobe with their sirens sounding noisily. Seeing my town like this was very painful. Then we began to do things, at first we cleaned my room. Then we went to see my friend's house. She had stayed in my house on the night of the earthquake, so she was very worried. We walked to her apartment house in Uozaki. On our way we called both our parents from a telephone booth. We had not been able to call from my house. When I heard my mother's voice, I cried. My parents also cried. I just said, 'Yes, I'm safe.' Many people waited in a long line to telephone, so I had to hang up quickly. Fortunately we saw my club mate's house safe. After cleaning her room, we returned to my apartment house. At noon we ate bread and drank milk in my room. My room was not broken, but part of my apartment house was broken. So we called into all the rooms again and again, 'Is there anyone there? If there is anyone, please answer.' But we were not able to hear anything at all. We weren't able to do any more. In fact, I'm afraid to say that my neighbour had been killed in the earthquake. The chimney of the next door public bath fell onto her room and killed her. But I didn't know this painful news at that moment. I did not know this news until I read an article in the newspaper. I feel bad, but I don't think we could have done any more – we thought that a big earthquake would shake again, and since my room would be dangerous, we escaped to my friend's safe room.

At dusk it was pitch dark in the room. There was no gas, no electricity and no water. We said 'All we can do is sleep.' But after a short sleep, we went to telephone again. While we were lining up, we were talking to a woman. She gave us good advice. 'Hey girls! Don't stay in house. Go to the shelter.' Then we went to Uozaki Junior High School. In the second night we couldn't sleep for cold and terror. Whenever an aftershock came, I gripped her hand firmly. I listened to the radio attentively. While the names of dead people were read out, I was praying that there was nobody I knew. I was putting myself first, I thought. But this is the truth. Before seven o'clock, a man said us 'There is an important news item, please listen.' The news was bad. We had to escape to the north area of Route 2 because of a gas leak. We forgot breakfast, we were in danger, so we left quickly.

At my club mate's house, I watched television. Television news was a greater shock than I had imagined. My friend and her family had been worried about me; they were very kind to me. But I hardly had time to relax. Her mother cooked lunch for me. It tastes very, very good – the first real food I ate in days. Then I called my parents. My father told me to come home to Kochi – he was worried too. I said goodbye to my friend and her kind family, and went to Kansai Airport. I went back to Kochi by plane. The moment I met my parents at Kochi Airport, I cried. I kept crying till we arrived at our home. But that is not the end. I am afraid of small aftershock yet. Whenever an aftershock happens, I think 'after this the big earthquake must come.' Now I am writing this Earthquake Experience, but I don't feel good. I can't forget. Perhaps this memory of the Hanshin Earthquake will follow me till my dying day. I will never get used to it and I will never forget. The sights of 17 January are still clear and alive.

1 Study Figures A and B.

  a) What are the main causes of death from volcanic eruptions?   (3)

  b) What is the meaning of the term *tsunami*?   (2)

  c) Why are there so many more casualties resulting from earthquakes than from volcanic eruptions?   (3)

  d) Complete Figure C. There were 141 earthquakes in 2001 and 133 in 2002. The number of deaths in 2001 was estimated as 21 357 and in 2002, 1660.   (2)

  e) Why was there such a large increase in deaths due to earthquakes in 1999 and in 2001?   (2)

Eruption	Year	Casualties	Major cause
Mount Pinatubo, Philippines	1991	900	Mudflows, roof collapse
Wum, Cameroon	1986	1 700	Asphyxiation by gas
Nevado del Ruiz, Colombia	1985	25 000	Mudflows
El Chichon, Mexico	1982	2 000	Ashflows
Mount St. Helens, Washington, USA	1980	57	Asphyxiation from ash
Mont Pelée, Martinique	1902	30 000	Pyroclastic flows
Krakatau, Indonesia	1883	36 000	Tsunami
Tambora, Indonesia	1815	92 000	Starvation
Unzen, Japan	1792	15 000	Volcano collapse, tsunami
Lakagigar (Laki), Iceland	1783	9 000	Starvation

A Deadliest volcanic eruptions since the 18th century.

Earthquake	Year	Casualties
Gujarat, India	2001	20 000
Turkey	1999	17 000
Afghanistan (two quakes)	1995	7 000
Kobe, Japan	1995	5 500
North-West Iran	1990	42 000
Northern Armenia	1988	50 000
Mexico City	1985	30 000
El Asnam, Algeria	1980	20 000
Tangshan, China	1976	650 000
Northern Peru	1970	70 000

B Deadliest earthquakes since 1970.

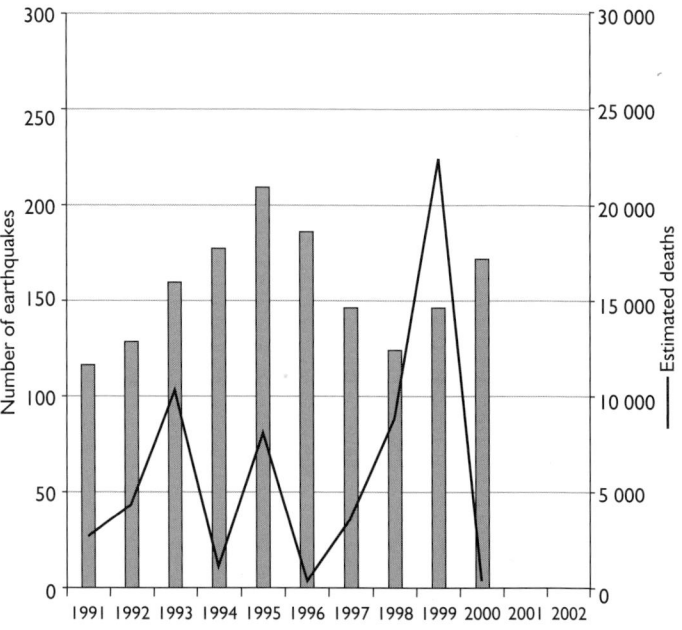

C Major earthquakes and estimated deaths 1991–2002.

2 Name the most deadly (i) volcanic eruption and (ii) earthquake of the twentieth century.   (2)

3 On an outline map of the world, mark the location of the volcanic eruptions and earthquakes listed in Figures A and B.   (10)

The Bhuj earthquake, which shook the Indian Province of Gujarat on the morning of 26 January 2001 (Republic Day), was one of the two most deadly earthquakes to strike India in its recorded history. Read the bare facts in the panel.

Date of occurrence:	26 January 2001
Time:	8:46 a.m.
Magnitude:	6.9 on the Richter scale
Epicentre:	23.6° North Latitude and 69.8° East Longitude, 20 km NE of Bhuj, 110 kms NNE of Jamnagar
Casualties:	• deaths 19 727
	• injured 166 000
	• homeless 600 000
Property losses:	348 000 houses destroyed and 844 000 damaged
Economic losses:	more than 20 000 cattle reported killed; estimates indicate total losses as high as $5 billion

1 Using the information in the panel (above), the news report and the eye-witness account on page 188, an atlas and Figures 16.1, 16.4 and 16.30 on pages 262, 263 and 272 of the pupil book, explain the likely cause of the Bhuj earthquake and describe the main effects. *(10)*

2 Explain why an earthquake in this remote area of India caused so many more casualties than the more severe earthquake which hit Kobe in Japan in 1995. *(5)*

3 Complete the comparison frame below in order to compare the effects of an earthquake in an LEDC and one in an MEDC. *(8)*

Effect	LEDC	MEDC
Buildings		
Casualties		
Transport		
Speed of human response		

The Bhuj Earthquake: early news report

In Ahmedabad city, the Shardaben hospital has reported 52 deaths, the civil hospital has reported 51 deaths, LG hospital 27, VS hospital 47 and Jivraj Mehta 10 deaths.

Many high-rises have collapsed in Ahmedabad city, the worst of them being the 10-storey Shikhar building. The worst damage is feared in the Ranip area of the city. At least 30 buildings are feared to have collapsed. In fact, so bad is the situation in the city, that Lakshmi hospital has run out of space and has taken to treating patients on the road outside.

The Morbi engineering college has crashed and 20 students are feared trapped in the debris. It is feared some 100 students of the Swami Narayan School at Mani Nagar in Ahmedabad are trapped under the debris of the school-building.

An Eye-witness Account from a Stricken Village

'I thought the earth had cracked open and was about to swallow me. It was the most horrifying experience of my life, don't ask me to narrate more. Right now I am sitting in the compound, my home has collapsed in the earthquake,' Vinubhai Sheth told rediff.com on the phone from his village, Khakhrichi, 100 km from Surat.

'I am a shopkeeper selling grains and other food items. My shop is also damaged. In just one day my world has been reduced to a shambles. Five people have died here, and we don't know if any more have died on the way to hospital.'

According to Sheth, some 80% of the buildings in his village, which has a population of 4000, have collapsed, and at least 25 people have been severely injured. They have been taken to hospital in Morbi, some 35 km away.

'Our village is a small one. Most of us are farmers, simple people, and unfailingly we attend the Republic Day function. That was what saved our lives today, all our children are safe because they were in the open for the flag-hoisting ceremony.'

Sheth himself was at home in the morning when he felt the earth move beneath his feet. 'I was shocked. I rushed to the road. When I looked back, my home was still shaking. Out of sheer fright I fell down and couldn't understand what was wrong with me. Like a child I crawled on to the road. After five minutes I got up and ran out of the village.' On the road there were many others like him. 'I saw all my neighbours running away. We were not just nervous, we were frightened. We saw death. My brother saw some three, four persons dead on the way. We are still scared. You have called from far away but the government is not around to ask about our welfare. There is no help available. People are hiring taxis to ferry the injured. In my village alone five people have died.'

Falguni Sheth, his niece told us, 'Our village has been reduced to dust. I was having breakfast when I got up to drink water and I felt the tremors. I don't know how to convince you about what we went through. Everywhere around me the destruction is total. Of course we are happy we are safe and alive.

'When the earthquake started I took my grandmother out of the house. We could not see a thing because the air was full of dust. Houses were collapsing and at one time it was so dark. I looked. I could not see my mother and grandmother for a few minutes, it was so scary. At present we are sitting in this compound, with just a telephone, since our relatives are calling us up from Bombay. We returned here to talk to them. All the buildings around us are broken, demolished or damaged beyond repair.'

The Eruption of Mount Pinatubo

Mount Pinatubo in the Philippines is one of a chain of volcanoes forming the Luzon volcanic island arc. The arc parallels the west coast of Luzon and reflects the subduction zone along the Manila trench to the west. The steep, easterly-dipping subduction of the Eurasian Plate started about 10 million years ago and produces a line of active volcanoes through the northern Philippines. In 1991, Mount Pinatubo was a dormant volcano – it had not erupted for over six hundred years. Yet it was the scene of the second-largest volcanic eruption of the 20th century and by far the largest eruption to affect a densely populated area. It provides an excellent example of human response to the threat posed by a natural hazard. The eruption occurred on 15 June 1991. There had been warning signs for two months before the violent eruption.

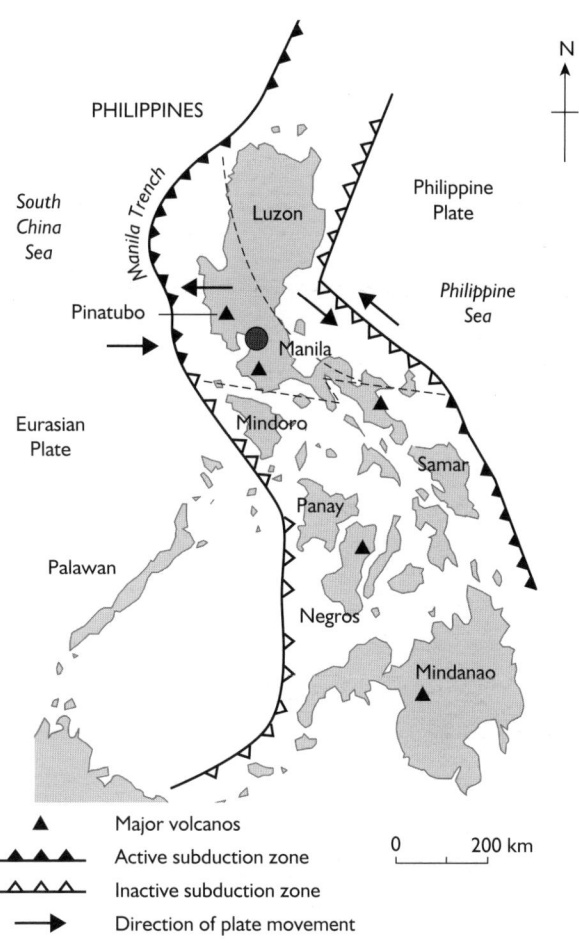

A The location of Mount Pinatubo, the Philippines.

▲ Major volcanos
▲▲▲ Active subduction zone
△△△ Inactive subduction zone
→ Direction of plate movement

0 ____ 200 km

Events leading up to the eruptions

16 July 1990: a powerful earthquake measuring 7.8 on the Richter scale struck about 100 km north-east of Mount Pinatubo, shaking the Earth's crust beneath the volcano. This major earthquake caused a landslide, some local earthquakes and a short-lived increase in steam emissions from a pre-existing geothermal area, but otherwise the volcano seemed to be continuing its 600-year-old slumber undisturbed.

March and April 1991: molten rock (magma) rising toward the surface from more than 32 km beneath Pinatubo triggered small earthquakes and caused powerful steam explosions that blasted three craters on the north flank of the volcano.

2 April 1991: people from the village of Patal Pinto saw small explosions followed by steaming and the smell of rotten eggs (hydrogen sulphide gas) coming from the upper slopes of Mount Pinatubo. Filipino and US geologists immediately installed portable seismometers near the

mountain and began recording several hundred earthquakes a day. Within two weeks the team had installed equipment capable of locating the increasing number of earthquakes. They began tape measurements across fractures opened during the early steam explosions and they later installed tiltmeters to detect new ground movement. With the help of the US Air Force, measurements of the sulphur dioxide content of the gas in the steam plumes, now continuously visible at Pinatubo, were begun. Using a scale of 1 (low-level events) to 5 (eruption), the geologists set alert level 2 on 13 May. Between 13 May and 28 May, a 10-fold

Alert level	State
1	Low level unrest
2	Unrest, probably involves magma movement
3	Eruption possible within two weeks
4	Explosive eruption possible within 24 hours
5	Eruption underway

rise in sulphur dioxide content was measured. All signals indicated that magma was rising within the volcano.

April, May and Early June 1991: Thousands of small earthquakes occurred beneath Pinatubo and many thousand tonnes of sulphur dioxide gas were also emitted by the volcano. On 5 June the geologists raised the alert to level 3 and on 7 June to level 4. Urged on by the geologists, evacuations of people began on 10 June. Two days later the first eruption occurred.

Pinatubo's eruption was ten times stronger than the eruption of Mount St. Helens and it threatened the lives of over one million people. More than 350 people died during the eruption, most of them from collapsing roofs. Many more would have died, but the Philippine authorities were able to evacuate 200 000 people from the slopes and valleys, and the American military evacuated 18 000 personnel and their dependents from Clark Air Base below the mountain.

Impacts of the Eruptions

Over 20 million tonnes of sulphur dioxide were injected into the stratosphere in Pinatubo's eruptions and dispersal of this gas cloud around the world caused global temperatures to drop temporarily, between 1991 and 1993, by about 0.5°C. The temperature drop was due to the sulphur dioxide combining with water to form droplets of sulphuric acid, which blocked some of the sunlight from reaching the Earth.

The Eruptions

- From 7 to 12 June, the first magma reached the surface of Mount Pinatubo. It oozed out to form a lava dome but did not cause an explosive eruption.

- On 12 June, millions of cubic yards of gas-charged magma reached the surface and exploded in the reawakening volcano's first spectacular eruption.

- When even more highly gas-charged magma reached Pinatubo's surface on 15 June, the volcano exploded in a tremendous eruption that ejected more than 5 cubic km of material. The ash cloud from this climactic eruption rose 35 km into the air. At lower altitudes, the ash was blown in all directions by the intense winds of a typhoon and winds at higher altitudes blew the ash south-westward. A blanket of volcanic ash and larger pumice pebbles blanketed the countryside. Fine ash fell as far away as the Indian Ocean and satellites tracked the ash cloud several times around the globe.

  Huge avalanches of searing hot ash, gas and pumice fragments (pyroclastic flows) roared down the flanks of Mount Pinatubo, filling once-deep valleys with fresh volcanic deposits as much as 200 m thick. The eruption removed so much magma and rock from below the volcano that the summit collapsed to form a large volcanic depression or caldera.

- By 16 June, when the weather cleared, the top of the volcano was gone, replaced by a 2.5 km wide caldera, and pyroclastic flow deposits had largely filled pre-existing valleys on all sectors of the volcano. Ash had fallen over a vast area beyond the volcano. The ash reached thicknesses of 30 cm as far as 40 km away from the volcano. The new summit height of Mount Pinatubo is approximately 1485 m above sea level, reduced from a pre-eruption elevation of 1745 m.

- Much weaker, but still spectacular eruptions of ash occurred occasionally through early September 1991. From July to October 1992, a lava dome was built in the new caldera as fresh magma rose from deep beneath Pinatubo.

The eruptions have dramatically changed the face of central Luzon, home to 3 million people. About 20 000 indigenous Aeta highlanders, who had lived on the slopes of the volcano, were completely displaced and many still wait in resettlement camps for the day when they can return home. About 200 000 people who evacuated from the lowlands surrounding Pinatubo before and during the eruptions soon returned home, but they faced continuing threats from lahars (giant mudflows of volcanic materials) that had already buried many towns and villages. Rice paddies and sugar cane fields that were not buried by lahars have recovered; those buried by lahars will be out of use for years to come.

Aircraft were warned of the dangers of the ash cloud. However, a number of jet airliners flying far to the west of the Philippines flew through the ash trail and sustained about $100 million in damage.

Continuing Hazards

Even after more than ten years, hazardous effects from the 15 June 1991 eruption of Mount Pinatubo continue. The thick, valley-filling pyroclastic-flow deposits from the eruption insulated themselves and have kept much of their heat. These deposits still had temperatures as high as 500°C in 2002 and may retain heat for decades. When water from streams or underground seepage comes in contact with these hot deposits, they explode and spread fine ash downwind.

Since the eruption, ash deposits have also been remobilised by monsoon and typhoon rains to form deadly lahars.

Exam Practice Questions

1 Draw a simple diagram explaining the cause of the Mount Pinatubo eruptions. *(4)*

2 What warning signs were there before the eruptions? *(3)*

3 What were (i) the immediate and (ii) the longer-term effects of the eruptions? *(6)*

4 a) How were people living close to Mount Pinatubo affected by the eruptions? *(2)*

b) How was a much greater human disaster avoided? *(4)*

The Eruption of Mount Pinatubo

These model answers are designed to show you how to get full marks for the longer questions in the Case Study Extra for this section. Read the question carefully and write your own answer. Then read the model answer and the examiner's notes to see what he or she is looking for when awarding the marks. Decide how many marks you think the examiner would give your answer. Decide how to change your answer to this particular question to increase your marks. What will you do differently next time you answer a similar question?

3 What were (i) the immediate and (ii) the longer-term effects of the eruptions? (6)

> Immediate effects included the explosive ejection of over 5 cubic km of material. The ash cloud rose 35 km into the air and travelled around the world several times before dispersing. Over $100 million of damage was caused to jet airliners that flew through the ash cloud. A blanket of fine ash and pumice covered the countryside up to 40 km around the volcano. Flows of hot ash and gas filled valleys to a depth of over 200 m.

> Longer-term effects include the fact that the eruptions blew the top off the volcano and created a 2.5 km wide caldera. The height of the volcano was reduced by 260 m. The gas cloud injected into the atmosphere caused global temperatures to fall by 0.5 degrees centigrade over the next three years. In the area around Mount Pinatubo giant mudflows are still triggered by heavy rains.

> **Level 1 (1–2 marks)**: *Basic answer including reference to lava flows.*
> **Level 2 (3–4 marks)**: *More detailed answer, including ashfall and its effects on the world's climate.*
> **Level 3 (5–6 marks)**: *Full, detailed answer.*

4 b) How was a much greater human disaster avoided? (4)

> Following warning signs in March and April 1991, scientists placed instruments on the volcano and monitored events. They issued warnings as the threat increased and finally urged the authorities to evacuate people living near the volcano. Evacuations began two days before the volcano erupted. This action saved thousands of lives.

> **Level 1 (1–2 marks)**: *Basic answer, including reference to evacuation of people.*
> **Level 2 (3–4 marks)**: *More detailed answer, including scientific monitoring of the volcano and issuing of warnings prompting evacuation.*

CHAPTER 17 covers arguably one of most important topics in Geography, on the basis that virtually all the Earth's surface consists of a network of river drainage basins. Moreover, rivers must be one of the most easily accessible fieldwork topics for all schools in the UK. This chapter consists of 10 short sections and a case study of the Yangtze, China. There are many photographs throughout, which would give opportunities for quizzes on the landforms, perhaps to conclude the chapter.

Drainage Basins   278 – 279

Drainage basins must be viewed as an open system in their own right as well as a subsystem of the hydrological cycle. Key terms and relationships are described here.

Activity Sheet 17.1 requires pupils to apply terms and concepts associated with drainage basins to a diagram provided.

River Discharge and Flood Hydrographs   280 – 281

The hydrograph is an extremely neat method of summarising flood events. All the necessary information can be tied together in one diagram, so enabling detailed comparison of different theoretical and real world situations. This section shows different basins with their hydrographs, linking the shape of the graph to the actual flood event on the ground. Annual hydrographs are also considered, with the River Torridge in Devon being used as an extended example.

Activity Sheet 17.2 also highlights the use of hydrographs, with definitions, a calculation of lag time and a link to the case study on Lynmouth on pages 286–287 of the pupil book.

River Processes   282

As a preliminary to river landforms, the key aspects of erosion, deposition and transportation of load are considered.

River Landforms in a Highland Area   283 and 292

This section includes a description of the features of a typical upland river valley.

Activity Sheet 17.3 supports the material on both pages 282 and 283 of the pupil book. There are multiple terms for some erosion processes, giving alternatives to suit teacher preference, as well as to familiarise pupils with the terms they may encounter in examination papers. Pupils are also required to describe a waterfall and young river

valley, as well as produce a field sketch, based on a photograph that ties in with the Case Study Extra on Pennine limestone landscapes in Chapter 15 of the *Teacher's Resource Book*.

River Landforms in a Lowland Area   284 – 285 and 292

This section includes descriptions of meanders, oxbow lakes, floodplains, levées and deltas. Page 292 in the pupil book provides examples of fluvial landforms, as well as extra skills practice.

Activity Sheet 17.4 concentrates on lower valley features with questions on flood plains, meanders and oxbow lakes. These questions are largely based on the interpretation of diagrams, as well as requiring pupils to produce their own.

Activity Sheet 17.5 focuses on deltas, considering physical characteristics such as shape and then human activities in these areas. The physical-human links are emphasised.

River Flooding in an MEDC and River Flooding in an LEDC   286 – 289

Case studies of the Lynmouth flood in 1952 and the regular problems in Bangladesh are presented. The Bangladesh situation is also briefly referred to in Activity Sheet 17.5.

Drainage Basin Mismanagement and Drainage Basin Management   290 – 291

The problems of waste and pollution are considered, followed by the introduction of the idea that it is not always the correct strategy to control a river's behaviour, although, there are times when this is beneficial.

Case Study 17 Flooding on the Yangtze – Management and Mismanagement and the Three Gorges Project   293 – 295

This controversial scheme is presented with photographs and consideration of its advantages and disadvantages. It provides an excellent opportunity for class debate.

The Improve Your Marks! section in the *Teacher's Resource Book* is based on this case study.

Case Study Extra Flood Prevention in Guildford

This is a small-scale study of basin management in the UK and acts as a contrast both to the mismanagement situations presented in the pupil book and the large scale of the Three Gorges Project, giving an overall balance to this topic.

1 What is meant by the term *hydrological cycle*. (3)

2 The hydrological cycle is a closed system. Explain what this means. (3)

3 a) What type of system is a drainage basin? (1)

   b) In what ways does this differ from a closed system? (3)

4 Define the terms (i) *inputs*, (ii) *outputs*, (iii) *stores* and (iv) *flows*, which are all characteristics
   of a system. (4)

5 On Figure A, complete the labels using the terms below. One has been done for you already.
   Label F represents an area that you will need to shade, both in the box and on the sketch map. (5)

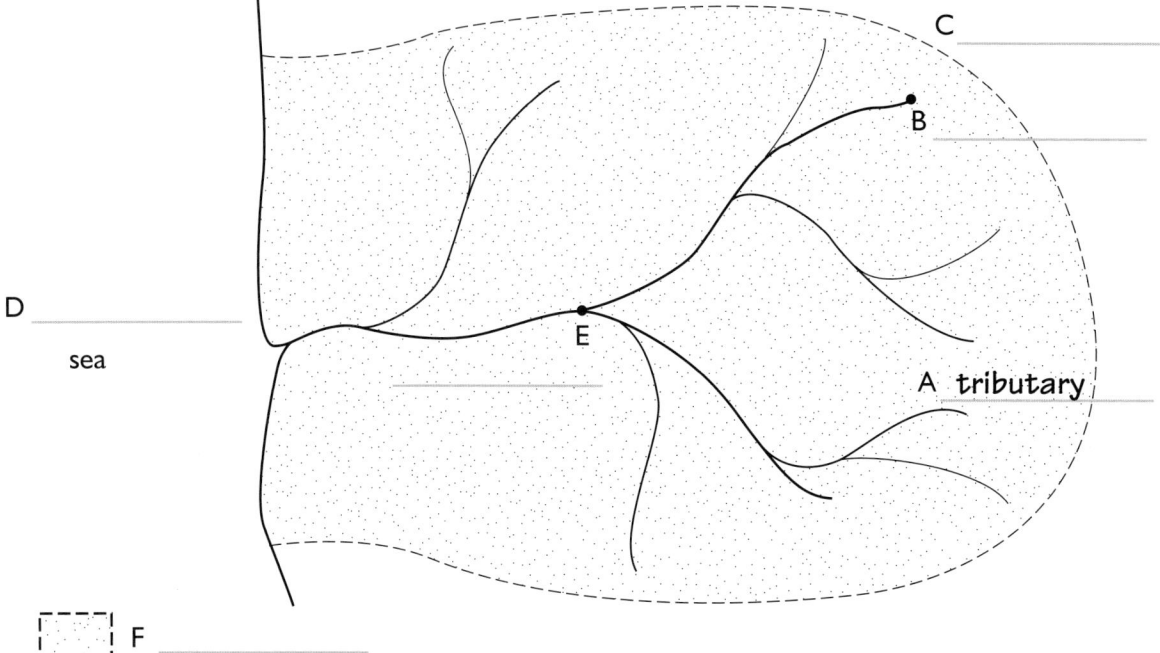

- Mouth
- Source
- Area drained by this river system
- Watershed
- Confluence

A A drainage basin.

6 a) On Figure A, add labels to show (i) the area of highest land, (ii) the area of lowest land
   and (iii) the area of moderate height. (3)

   b) Shade each area an appropriate colour. (3)

   c) Add a key in the space available to explain your shading. (3)

7 Describe what happens to rain falling within this basin

   a) if the rock type is impermeable (3)

   b) if the rock type is permeable. (3)

1 Write each of the following terms from the panel in the correct place, A–E, on Figure A. (5)

- Peak discharge
- Lag time
- Falling limb
- Peak rainfall
- Rising limb

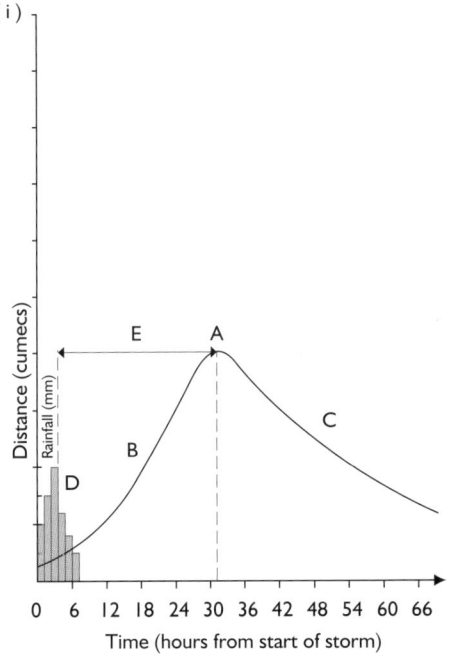
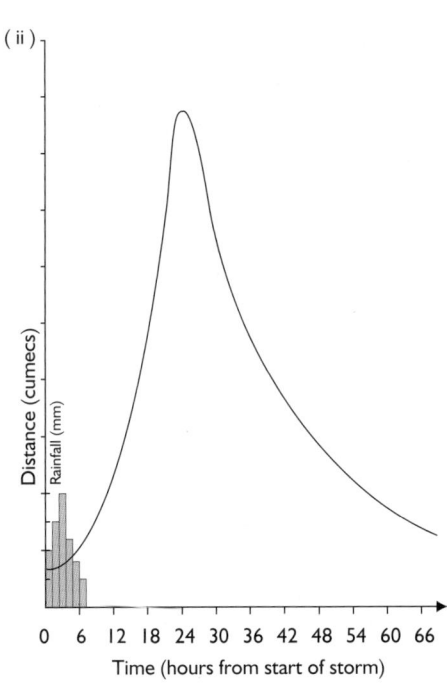

A Hydrographs (i) X and (ii) Y.

2 a) Calculate the lag time for hydrographs X and Y in hours and minutes. (4)

b) Which hydrograph has the longer lag time? (1)

c) Explain the importance of lag time in a flood event. (3)

3 a) Explain the term *surface runoff*. (2)

b) Which characteristic of a drainage basin most affects the steepness of the rising limb? Explain why this is so. (3)

4 a) Define the term *drainage density*. (2)

b) How is drainage density calculated? (2)

c) How does drainage density affect flooding? (4)

5 Discuss the ways in which different land uses affect the likelihood of flooding in a drainage basin. Refer to hydrographs X and Y to help you explain. (9)

6 Read the case study of the Lynmouth flood of 1952 on pages 286 and 287 of the pupil book.

a) Which of the two hydrographs on this activity sheet, X or Y, would fit the Lynmouth flood better? (1)

b) Give two reasons for your answer. (4)

1 a) Describe (i) abrasion/corrasion and (ii) attrition, which are two types of river erosion. (4)

b) Explain the relationship between abrasion/corrasion and attrition. (2)

c) Name two other ways in which rivers can erode. Choose one of these and describe it in detail. (5)

2 Rivers carry their load in four main ways.

a) Name the four main methods in which a river transports its load and briefly explain each one. (8)

b) Draw a diagram to show any one of these transportation methods. (3)

c) Another, less important, way in which rivers transport loads is known as floatation. What, sorts of materials are carried this way and how do they originally get into the river? (3)

3 Figure A shows an upland river in the Yorkshire Dales National Park in the Pennines.

a) Draw a field sketch, based on the photograph, and label it to show (i) landforms caused by river erosion and (ii) evidence of transportation by the river. *(5 for sketch + 5 for labels)*

b) Choose one landform labelled on your sketch.

(i) Describe its appearance. (3)

(ii) Explain its formation. (4)

A Thornton Force, Yorkshire Dales.

1 Imagine you were on a walk along a small river from its source to its mouth. Complete the table below with your comments on how the characteristics of the valley would change as you progressed downstream. The first one has been done for you. *(6)*

Feature	Upper valley	Lower valley
Width of channel	Narrow	Wide
Width of valley floor		
Height of valley floor above sea level		
Straightness of valley sides		

2 Explain the formation of a flood plain in a river valley. *(5)*

3 Meanders begin when a river is forced to change direction slightly. Erosion is then uneven across the channel.

a) On Figure A(i), shade all the places where erosion and deposition occur. *(5 + 5)*

b) Add a key to explain your use of colour. *(2)*

c) Figure A(ii)–(iv) shows three channel cross sections. Label each of the locations marked on Figure A(i) with (ii), (iii) or (iv) to indicate the position of each cross section. *(3)*

d) On Figure A(i), label one river cliff, one slip-off slope, a meander core and a meander neck. *(4)*

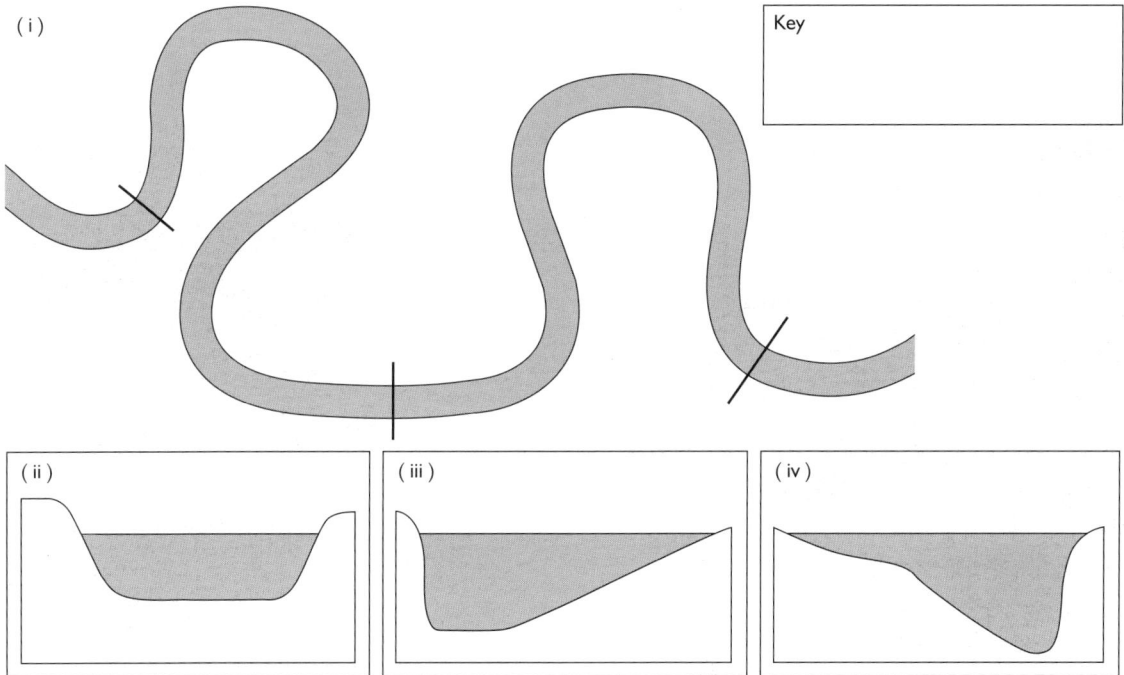

A (i) A meandering river with three channel cross sections, (ii)–(iv).

4 With continued erosion, meanders can develop into another feature.

a) Name this feature. *(1)*

b) Draw a labelled sketch to show the formation of the new feature. *(5)*

c) What happens to the shape of the main channel after the formation of this new feature? *(2)*

1 a) Study Figure A and describe the shape of the Nile delta. *(3)*

  b) There are two main shapes of delta. Give their names. *(2)*

  c) Explain why these two shapes of delta occur. *(4)*

2 Look at Figure 17.20 on page 285 of the pupil book, which shows the delta of the River Mississippi.

  a) Use your atlas to locate the Mississippi delta. Name the state and the country it is in, and the sea into which it flows. *(3)*

  b) In what ways is the delta of the Mississippi different to that of the Nile? *(4)*

3 What is the difference between a tributary and a distributary? *(2)*

4 Deltas are often used for various human activities. They have several advantages, but also some disadvantages.

  a) Which physical characteristics of a delta make it good for farming? *(3)*

  b) Which physical characteristic of a delta can make it difficult for farming? *(1)*

  c) Use your answers to a) and b) and the information in the panel below to explain the success of Egyptian agriculture in the Nile delta region. *(5)*

Mediterranean Sea

N

Alexandria

Cairo

Red Sea

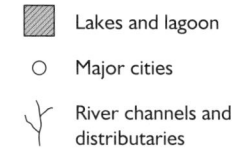

▨ Seas

▨ Lakes and lagoon

○ Major cities

⅄ River channels and distributaries

☐ Main agricultural areas

A The Nile delta, Egypt.

The main crops produced in Egypt are cotton, wheat, rice, sugar, fruit, vegetables and beerseem (a type of clover grown as a fodder crop). Egypt is one of the world's biggest cotton growers. Many people are employed in the textile industry (spinning, weaving and dyeing of fine quality cotton) and in clothing manufacture. Also, Egypt is the world's largest producer of dates, most of which are exported.

5 a) The main part of the River Mississippi is a very important transport route through the eastern half of the country. Use Figure 17.20 in the pupil book to explain what transport difficulties the delta may cause. *(3)*

  b) How can such difficulties be overcome? *(2)*

6 Research the physical characteristics of other major world deltas and the economic activities that take place there, for example the Ganges/Brahmaputra delta in Bangladesh or the Rhine delta in the Netherlands. Present your findings to your group or class.

Flood Prevention in Guildford

In the UK at the present time, some settlements are relatively well protected from the flood risk affecting them, while others seem to be suffering increasingly often. Some problem areas have defences that were built before the 1990s, when the risk of floods became greater than in previous years. Settlements such as Yalding in Kent are now being flooded regularly. However, this case study looks at the Guildford town centre flood management scheme, which has worked very well so far. Future plans for other settlements might usefully be based on this scheme.

Figure A shows the location of Guildford and the River Wey. Guildford town centre is situated in a gap in the North Downs and the valley at that point is particularly narrow. The town is on one of the few transport routes to pass through the Downs and suffers from the resulting congestion. Much of the town, especially the residential areas, sprawl up the valley sides, making for some steep streets, including the High Street. In general, the narrow valley floor was not built on, except for purposes like car parking that would sustain little damage in a flood.

Plans for Development

During the 1960s and 1970s, the central business district (CBD) grew outwards onto the flood plain. New developments that needed a considerable amount of land were in demand. It was important that their locations were within easy reach of the CBD, where the availability of land was already restricted by existing developments. The new developments included:

• the Yvonne Arnaud theatre

• a large department store

• new units for the industrial estate.

In planning these new developments, the perennial problem of the flooding of the River Wey as it flowed through the town centre had to be taken into account. There was obviously little point in building on sites where flooding was inevitable without also taking precautions to reduce the risk of flooding. Figure B on page 200 maps the locations of the wide range of small-scale flood defences that were constructed on the floor of the River Wey valley. Figure C on page 200 and Figure D on page 201 shows some of them photographically. The descriptions that follow relate to the numbered sites on Figure B.

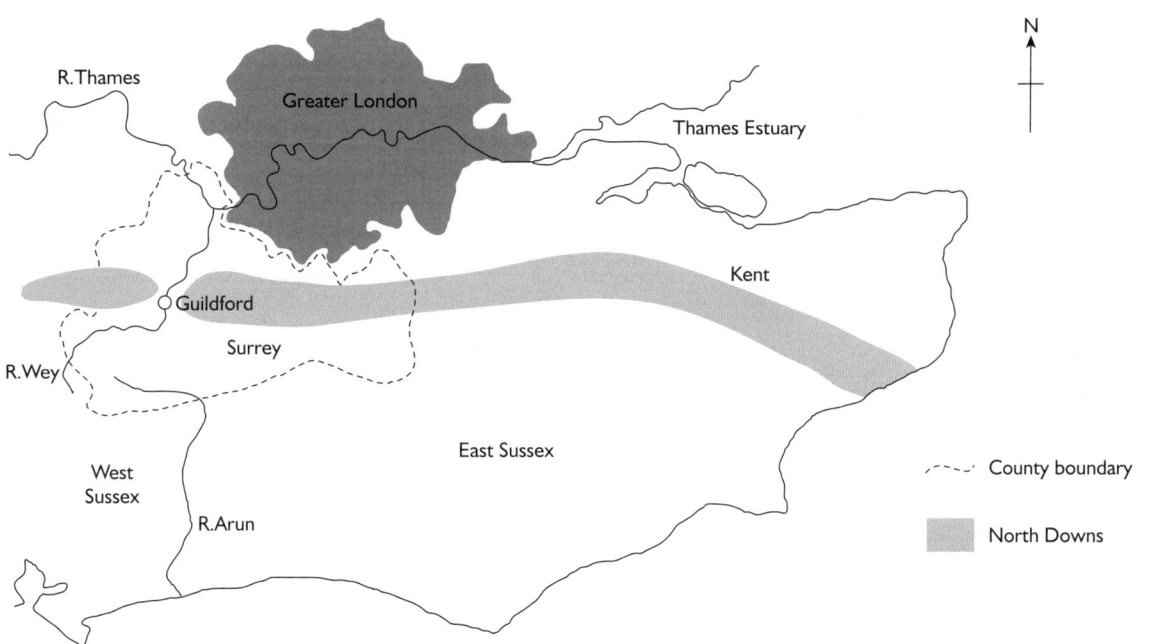

A Guildford and the River Wey. Guildford lies on the River Wey, a tributary of the River Thames. The Wey rises in the central part of the High Weald in south-east England. It has been used as a line of transport between the Thames valley and the south coast for many years and a canal link was built between the River Arun in West Sussex and the Wey basin.

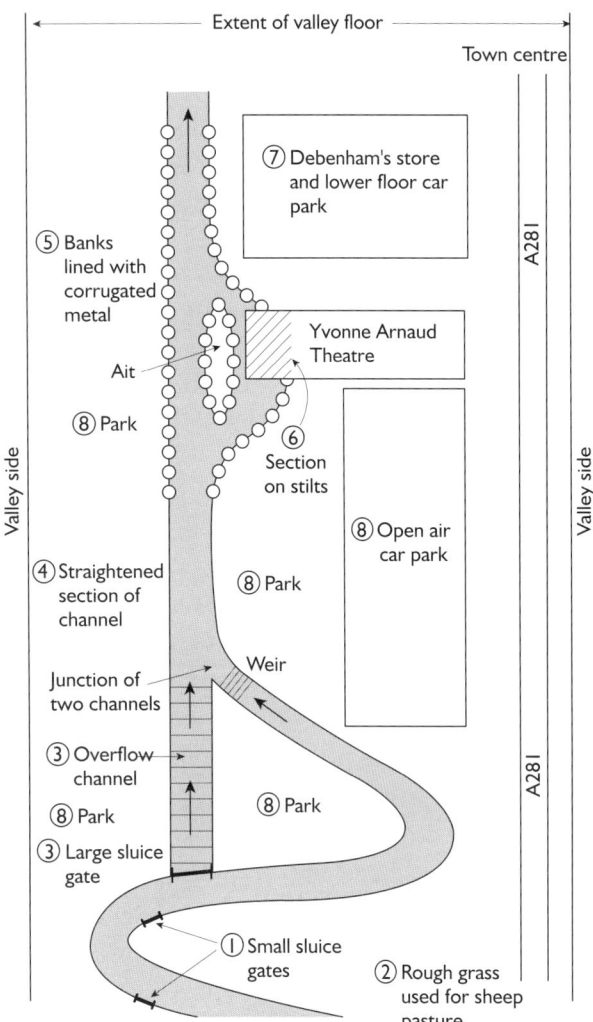

B Simplified sketch to show Guildford flood defences in the floor of the Wey valley.

C Part of Guildford flood defences, a large sluice gate, marked 3 on Figure B.

Flood defences

1 Sluice gates in the levées upstream of the town centre

Small natural levée development has been reinforced to make more useful continuous banks. To a considerable extent these help contain raised water levels in the channel. However, when river levels are particularly high and threaten to flood over the levées, small sluice gates, built into the levées at regular points, are opened.

2 Sheep grazing on the flood plain upstream of town

Flood water from the opened sluices is allowed to spill out onto the flood plain at chosen locations in a controlled way. River levels are therefore reduced before passing through the town centre. Instead of allowing flooding to occur in the town centre, where damage would be much more costly, water is allowed to cover the rough pasture just upstream. Any sheep can be moved elsewhere before this occurs.

3 The overflow channel

Should the diversion of water onto the flood plain not be sufficient, a larger sluice gate can be used. This effectively splits the channel into two, providing at least double the water capacity, and helps to speed the flow of the water through the town centre. The amount of water entering the overflow channel can be controlled.

4 Straightened channel

From the point at which the overflow channel rejoins the natural one, the speed of flow is quickened in two ways. One of the methods was to straighten the channel. Without meanders to move around, the water is not slowed by friction to such a great extent and so flow is more efficient.

5 Channel lined with corrugated metal

Along the straightened section, the efficiency of the channel is increased still further by lining the banks with corrugated metal. Smoothing the channel sides in this way is a further method of reducing friction.

6 Construction on stilts

The Yvonne Arnaud was planned as an important regional theatre and so had to be quite large. However, building it on the flood plain meant there was not much space. The rear part of the building was therefore constructed on stilts and it actually overhangs a quiet, wider part of the river. This means that the floor height is above past flood levels, which should protect it from damage should there be a future flood.

7 Car park on ground floor of department store

When the department store was built in the 1970s, space for such a store in the CBD was not really available unless this risky site was utilised. Therefore it was decided to build upwards to five storeys, but use the lower areas for parking. The parking was also an obvious attraction for potential customers and so benefited the business.

8 Low risk use of other locations in the valley floor

The remainder of the flood plain in the town centre section is used for purposes that will not suffer much in the event of a flood, such as car parking, park and play areas. In fact it is a benefit to have some open green space in the middle of the town as well as to have additional parking.

Conclusion

Guildford is an example of the use of several small-scale flood prevention measures in one small area. The whole programme has been effective while remaining attractive; indeed, in terms of land use, the area is an asset to the town.

D Channel lined with man-made materials (downstream from the section shown on Figure B).

Exam Practice Questions

1 State five ways in which people use river valleys that increase the risk of flooding. *(5)*

2 Suggest some reasons for the increasing risk of flooding in the world today. *(4)*

3 Select a flood prevention scheme you have studied. Describe the scheme and discuss its advantages and disadvantages. You could use the Guildford case study, the Three Gorges Project on pages 294–295 of the pupil book or another area you know from your own work. *(9)*

Flood Prevention in Guildford

These model answers are designed to show you how to get full marks for the longer questions in the Case Study Extra for this section. Read the question carefully and write your own answer. Then read the model answer and the examiner's notes to see what he or she is looking for when awarding the marks. Decide how many marks you think the examiner would give your answer. Decide how to change your answer to this particular question to increase your marks. What will you do differently next time you answer a similar question?

1 State five ways in which people use river valleys that increase the risk of flooding.   (5)

- Car parks, sports areas and any other uses that involve concreting over large areas of ground.

- New buildings of any sort, because these are impermeable surfaces:
    - public buildings
    - housing estates.

- Drains for built up areas; water movement to rivers is speeded up.

- Transport routes – roads, railways and airports all involve hard surfaces and drains.

This is not a 'Levels' question, even though it carries five marks. You are asked just to state uses that increase risk, so no extra detail or explanation is needed. It may be tempting to add this in if you know it, but don't be tempted! By doing so you will be wasting time that will be valuable for other sections. You could give your answer as bullet points, which makes your ideas really stand out to the examiner. It is quite easy to think of three or four answers, but a fifth takes rather more ingenuity!

---

3 Select a flood prevention scheme you have studied. Describe the scheme and discuss its advantages and disadvantages.   (9)

The Three Gorges Project is a large multipurpose scheme. In other words it solves several problems at once, of which flooding is only one. Flooding downstream from the development site should end because the huge reservoir enables the summer floodwater to be stored and then released in a controlled way later in the dry winter months. Problems caused by almost drought conditions in some winters will therefore also be a thing of the past. Millions of people will be saved from this huge risk. In the past floods along the River Yangtze have taken millions of lives.

However, a major disadvantage of this scheme is that a huge area upstream will inevitably be flooded behind the dam. The depth of the new lake will be greater than the height of medium-rise flats today. Hundreds of thousands of people will lose their homes and land and will be forced to migrate and start again elsewhere. They will, of course, have new modern homes provided with better facilities.

The shortage of electricity in this rapidly industrialising country will also be addressed by the hydro-electric power provided by the dam, where two huge power stations are part of the project. More jobs will also be created as a result, increasing the average standard of living. Moreover, hydro-electricity is a clean and renewable form of energy, so there is less pollution and consequence to people's health. However, water pollution may result when factories are flooded behind the dam.

The Yangtze is a major transport route. A lock and a lift for ships have both been incorporated in the project. This will allow larger ships to travel further inland more safely. Tourism will also benefit. However, the impact of the huge lake will result in the winter climate becoming milder and, unfortunately, wetter.

The greatest risk involved in the project is its size – nothing so large has ever been attempted before and so the potential impact on the surrounding natural environment is unknown.

Improve Your Mark!

Although this answer is based on the case study of the Three Gorges Project on pages 293–295 of the pupil book, it could have been based on the Guildford case study. The projects are on very different scales, but both provide enough material for a good answer to the question.

The key thing to remember when a question asks for advantages and disadvantages is that some sort of balance between the two must be achieved. There are nine marks allocated for the question, so at least four or five good points need to be made on each side of the argument. You must also remember that with 'Levels' marking it is the quality of your discussion that counts and not just the number of points you make.

Level 1 (1–3 marks): *At this level your answer would either include only a short description, followed by one or two advantages and disadvantages, or include mostly description and not really explore the advantages and disadvantages.*

Level 2 (4–6 marks): *In this case, the description would give more detail, for example that the project is multipurpose, the numbers of people involved and how their lives would be improved; two or three advantages and disadvantages would also be given.*

Level 3 (7–9 marks): *For a really top answer, the initial description would be thorough. The model answer contains eight advantages and five disadvantages, which is plenty. Go through the answer and note where these points have been made.*

CHAPTER 18 is divided into seven sections: coastal processes and erosion landforms, coastal processes and deposition landforms, causes, effects and responses to cliff erosion, coastal landforms on OS maps, the need for coastal management, coastal flooding in MEDCs and coastal flooding in LEDCs. Case studies within the chapter include the coast of the Isle of Purbeck, the New Forest coastline, the 1953 storm surge in England and the Netherlands and coastal flooding in Bangladesh.

Coastal Processes and Erosion Landforms 📖 300 – 301

This section introduces the relevant terms and concepts to do with waves, coastal erosion and the resultant landforms, including headlands, bays and the cliff to wave-cut platform sequence.

Activity Sheet 18.1 covers the processes of coastal erosion and includes a map of the coastline at Woolacombe in North Devon to test pupils' understanding of differential erosion.

Activity Sheet 18.2 focuses on Marsden Rock, which is an imposing 100-foot sea stack of magnesian limestone lying approximately 100 yd off the mainland of South Shields. Until 1996 it featured an impressive natural arch. However, in 1996 the top section of the arch collapsed, splitting the rock into two separate stacks. In 1997, following the relentless battering of the North Sea, experts inspected the smaller stack to the south of the main rock and declared that it was so unstable that it presented a danger of collapsing onto the shore. In the interest of public safety, the decision was made to demolish it.

Activity Sheet 18.3 presents a study of the collapse of the Holbeck Hall Hotel in Scarborough in 1993. It includes information and questions on the coastal management process of managed realignment.

Coastal Processes and Deposition Landforms 📖 302 – 303

This section introduces the terms and concepts relevant to transportation by longshore drift, coastal deposition and the resultant landforms including beaches and spits. Sea level changes are also introduced, with definitions of fjord and rias.

Activity Sheet 18.4 involves use of the website www.multimap.com to find a map of Hengistbury Head in Dorset. This provides a map and an aerial photograph of the headland featuring a large concrete groyne, which illustrates the consequences of the up and down-drift of groyne construction well. Note the small groynes along Mudeford Spit. Before the groyne was built in 1938, Mudeford Spit was much longer and wider, as the website

www.old-maps.co.uk will show. The groyne starved the spit and many groynes had to be built on Mudeford Spit itself – with only limited success.

Activity Sheet 18.5 presents questions on bars and tombolos, based on information provided on Slapton Sands, Devon; St. Ninian's Ayre, Shetland; and Chesil Beach, Dorset.

Causes, Effects and Responses to Cliff Erosion 📖 304

This page includes some methods of hard and soft coastal protection.

Coastal Landforms on OS Maps 📖 305

This page includes a 1:50 000 OS map of the eastern coast of the Isle of Purbeck, with a geological map. An OS map of Hurst Castle Spit is also included.

The Need for Coastal Management 📖 306 – 309

An extended study of the coastline of the New Forest deals with the pressures facing the coast and its main management problems. There follows a study of the coastal management plan, its aims usefully summarised in Figure 18.25 on page 308.

Case Studies 18a and 18b Coastal Flooding in MEDCs and in an LEDC 📖 310 – 313

The 1953 North Sea floods form the basis of this case study. The concept of the storm surge is introduced and studied in detail. The disastrous effects of the flooding led to major flood protection schemes in England and the Netherlands, including the Thames Flood Barrier.

Flooding in Bangladesh is the subject of this case study. Repeated storm surges have brought disaster and hundreds of thousands of deaths. Suggested policies to reduce the impact are outlined.

Case Study Extra Protecting Mappleton

Mappleton is a village on the coast of East Yorkshire, which provides an excellent case study on the implications of human interference with the coastal system. The desire to protect the village from coastal erosion led to the construction of an expensive hard engineering scheme, which has succeeded in halting the erosion at Mappleton (at least for a few decades), but at the expense of the coast down-drift from Mappleton's new defences. The resultant loss of a farmhouse hit the national headlines. There are excellent photographs of Mappleton at the following website:
www.atschool.eduweb.co.uk/nelthorp/room8/intra/geogra ph/humberside/holderness

1 The following diagram gives definitions for four coastal processes. Write the letter for each definition under the correct process in the table below.   (4)

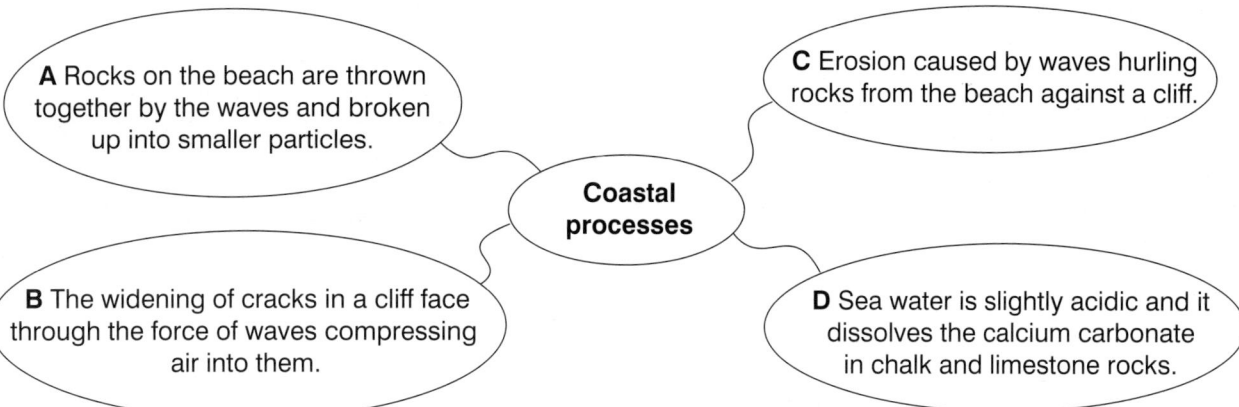

A Rocks on the beach are thrown together by the waves and broken up into smaller particles.

C Erosion caused by waves hurling rocks from the beach against a cliff.

Coastal processes

B The widening of cracks in a cliff face through the force of waves compressing air into them.

D Sea water is slightly acidic and it dissolves the calcium carbonate in chalk and limestone rocks.

Process	Corrasion	Attrition	Solution	Hydraulic pressure
Definition A,B,C orD				

2 Study Figure A. How has the series of bays and headlands along this section of the coast of North Devon been formed?   (5)

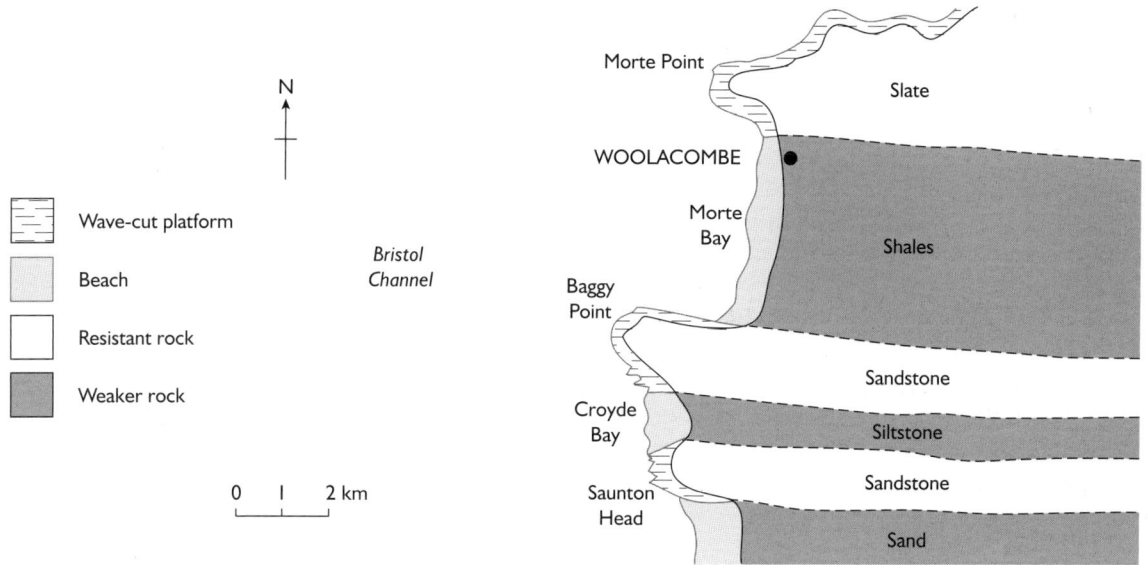

A The coastline at Woolacombe in North Devon.

3 Draw a labelled diagram to explain the formation of a wave-cut platform.   (6)

1 Study Figures A and B, which are field sketches of Marsden Rock, which lies off the coast of north-east England near South Shields. Note that the sketches are separated in time by ten years.

   a) Name the landforms marked X and Y on Figure A. *(2)*

   b) Name the landform marked Z on Figure B. *(1)*

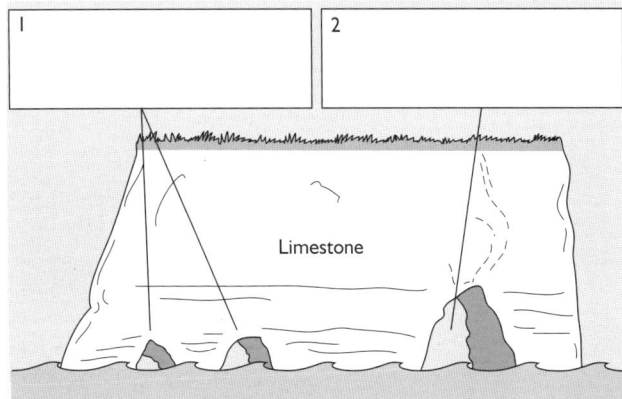

A Marsden Rock in 1986.

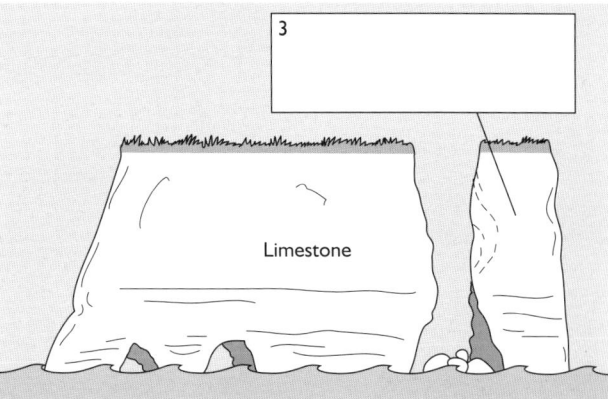

B Marsden Rock in 1996.

2 Describe and explain how this stretch of coastline changed between 1986 and 1996, using diagrams to help you. *(8)*

3 What is likely to happen to Marsden Rock in the future? *(2)*

4 Read the information about coastal protection methods in the panel. Then read the statements about sea defences in the table. Write the name of each sea defence next to the correct statement in the table. *(4)*

Coastal protection methods

Sea wall

Constructed of stone or concrete, sea walls usually have a curved top to deflect wave energy back onto the next incoming wave.

Rock armour

A collection of large interlocking boulders placed to protect the coast by breaking up the waves before they can hit the cliff foot.

Groynes

A long breakwater, made of wood or concrete, built out into the sea at right angles to the shore. Groynes are designed to stop longshore drift and build up the beach in order to protect the cliff foot.

Statement	Name
The coast beyond may not have enough beach material and so erosion occurs.	
Designed to reflect wave energy.	
Stops longshore drift and keeps beach material in place.	
Effective in breaking up the waves before they hit the cliff foot, but can look ugly on a beach.	

1 Study the information in the panel, which describes the fate of the Holbeck Hall Hotel in Scarborough.

a) Describe and explain the causes of the cliff collapse. *(5)*

b) (i) Why were people taken by surprise when the collapse occurred? *(2)*

   (ii) Why were there no casualties from the collapse of the hotel? *(1)*

The Collapse of the Holbeck Hall Hotel

The Holbeck Hall Hotel was the only four star hotel in Scarborough, North Yorkshire. Built in 1881, the hotel occupied a superb site 65 m above sea level in the South Cliff area of the town, with panoramic views over the town. In front of the hotel an extensive lawn ran down to the cliff top, a fairly gentle, grassy slope criss-crossed with footpaths, with a much steeper face in the final 12 m drop to sea level where there was a stone-built promenade. For 112 years, the Holbeck Hall graced the Scarborough coastline.

At about 7.30 a.m. on Friday, 4 June 1993, a major landslide occurred. A guest at the hotel said 'I couldn't believe what was happening. The landscape had changed so dramatically. The whole thing looked dreadful. The footpaths were at all different angles.' A 700 m stretch of cliff top had collapsed, removing two-thirds of the hotel's lawn. Eighty guests and staff were evacuated at 8.00 a.m. At 3.30 p.m. the same day, a second landslip removed the rest of the lawn and the hotel's

conservatory tumbled over the edge. On Friday evening and Saturday morning a third slip occurred. The following evening, the north-east wing of the hotel collapsed as a fourth slip undermined the hotel. Only half of the hotel remained when the landslips finished; this was declared unsafe and had to be demolished.

The coastline at South Cliff, Scarborough is composed of glacial boulder clay. A series of dry years in the early 1990s caused the clay to dry out and crack. Heavy rainfall in April and May 1993 then saturated the clay. The weight of the hotel caused the cliff to become unstable and it slumped along a curved plane (rotational slip). Such rotational slips are common along the Yorkshire coast, but the scale of the slip that claimed the Holbeck Hall Hotel was unusual. Erosion by the sea played little part in this disaster, but elsewhere erosion of the base of the cliffs leads to steeper, less stable slopes, which are more likely to slip.

---

2 Name and explain two methods used to protect cliffs from collapse. *(4)*

3 Some people are in favour of coastal defences and other people are against them. Give reasons for both viewpoints. *(4)*

4 Read the information in the panel on the process of managed realignment. Now explain the process in your own words. What advantages and disadvantages does this form of coastal protection have? *(4)*

The Process of Managed Re-Alignment

A new form of coastal protection known as managed re-alignment works with nature instead of trying to control it. Existing seawalls built on the shoreline are breached to allow seawater to flood the land behind, creating salt marshes. New seawalls are built further inland behind the saltmarsh. The largest managed realignment in the UK, covers an 8 km stretch of the coast of Lincolnshire, at Frieston Shore near the town of Boston. Three 50 m breaches have been cut in the outer sea bank, letting salt water from The Wash encroach on

78 hectares (193 acres) of former farmland purchased by the Royal Society for the Protection of Birds.

The marsh attracts birds and other creatures, besides protecting the coast more cheaply than artificial defences. Record numbers of avocets have already nested on the marsh, the first to breed there for a century.

The saltmarsh absorbs the energy of the waves. The cost of building a sea wall is 10 times less if you have a saltmarsh in front of it than if it faces straight onto the incoming tide.

---

1 Study Figure A.

a) Name the process shown in this diagram. *(1)*

b) Describe and explain the different appearance of the beach at X and Y. *(3)*

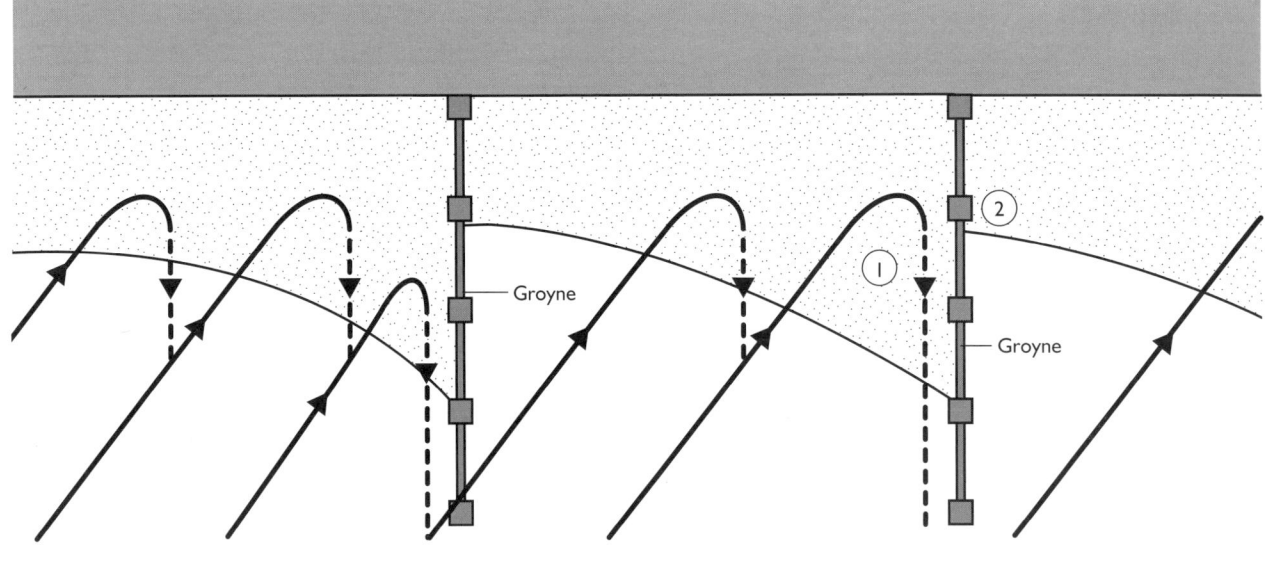

SEA

A Wave action on a beach.

2 a) Visit the website www.multimap.com to find a map of Hengistbury Head in Dorset. Type Hengistbury Head into the 'QuickSearch GB Postcode or Place' box on the home page. Then click on the button to access an aerial photograph of the headland. Print the photograph. Describe and explain the appearance of the beach to the west and to the east of the large groyne. *(5)*

b) On the same website map, move north one mouse click using the arrows shown in the left margin. You will observe a narrow strip of land running in a northerly direction.

(i) Name this landform. *(1)*

(ii) How has this landform been formed? *(2)*

(iii) What measures have been taken to protect this landform? *(2)*

(iv) Why do you think such measures are necessary here? *(3)*

Longshore drift creates other landforms in addition to beaches and spits.

1 Study Figures A and B.

    a) With the aid of labelled diagrams, explain how (i) bars and (ii) tombolos are formed.   (8)

    b) Why is the lagoon that forms behind a bar only a temporary feature?   (2)

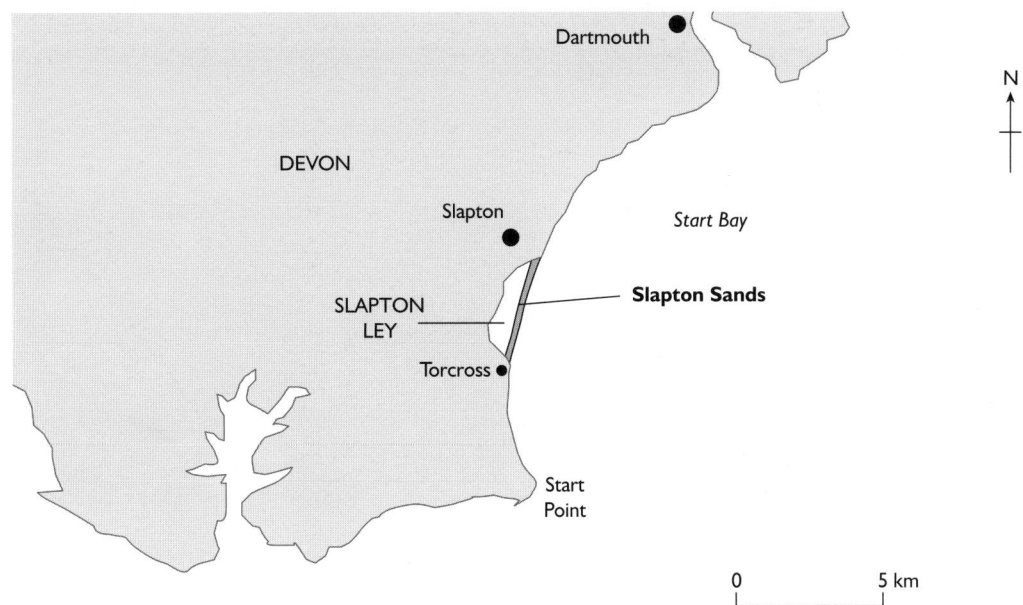

A Bar at Slapton Sands, South Devon. In some places a spit may grow right across a bay or estuary to form a bar. The stretch of water dammed up behind the bar is called a lagoon. The lagoon is often only a temporary feature. It is soon filled in by material deposited there by waves breaking over the bar or by rivers flowing into the lagoon. Slapton Sands in South Devon is an example formed of flint shingle, which has dammed up the lagoon of Slapton Ley.

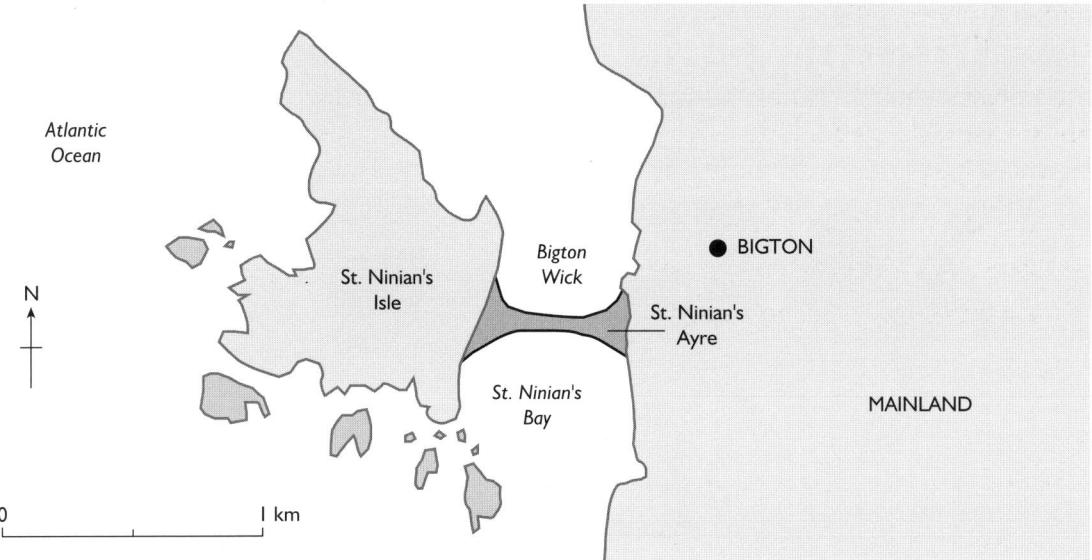

B Tombolo, St. Ninian's Bay, Shetland Isles. Sometimes a spit may grow out into the sea until it reaches an island, which it then links to the mainland. In this case the landform is called a tombolo. One of the best examples in the British Isles is St. Ninian's Ayre in the Shetland Isles, a 400 m long tombolo of sand linking the Isle of Mainland to St. Ninian's Isle. It has been built up by wave action from opposing directions, bent around (or refracted) in the shelter of the island. At each end of the tombolo are sand dunes. Photographs can be viewed at the following websites: **www.hp.europe.de/kd-europtravel/shetland/ stninian.htm www.shetland-heritage.co.uk /brochures/area_pages/ south_mainland /st_ninians_isle.htm** and **www.orkneyshetland.co.uk/shetland.html.**

2 What alternative explanation is there for the formation of Slapton Sands in Figure A and Chesil Beach in Figure C?

(4)

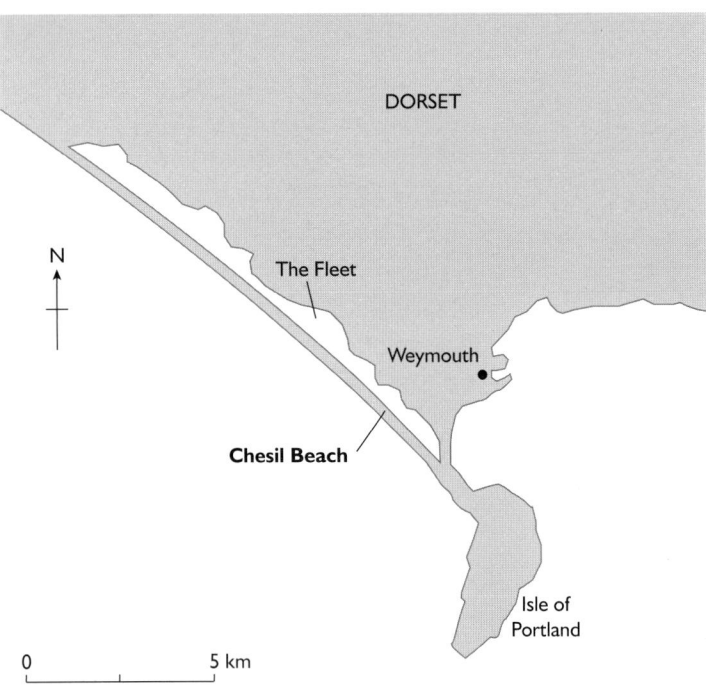

C Chesil Beach, Dorset.

An Alternative Explanation for Bar and Tombolo Formation

Slapton Sands in South Devon (Figure A) is often given as an example of a bar. It certainly looks like one, but some geographers think that this landform was actually formed in a different way. They state that the lagoon, Slapton Ley, was created by the natural damming of an estuary by a shingle bar, not as the result of longshore drift, but through being pushed up by the rising sea-level during the 10 000 years since the last Ice Age. This theory is supported by the discovery that the freshwater lagoon itself is about 3000 years old, certainly not a temporary feature. The surface layers of the shingle forming the bar have been sorted by longshore drift, but longshore drift was not responsible for its formation.

A similar process may explain the formation of Chesil Bank in Dorset (Figure C), itself often quoted as an example of a tombolo.

Protecting Mappleton

Mappleton is a village of just over one hundred people, 3 km south of Hornsea in the Holderness district of East Yorkshire. It is the site of a controversial coastal protection scheme. This is the fastest retreating coastline in the world; since Roman times the coastline has retreated by an average of 6km. Dozens of villages and towns have been lost during this process (see Figure A).

The Reasons for Rapid Erosion

■ The plain of Holderness did not exist before the Ice Age. It was a wide bay backed by chalk cliffs running from Flamborough Head to Hessle, west of the city of Hull as shown on Figure A.

■ Today Holderness is made up of glacial tills – sands and clays deposited by ice sheets during the Ice Age about 18 000 years ago. The tills are soft, unstable and have little resistance to erosion.

■ The low cliffs repeatedly slump down along rotational slip planes, lubricated by water which reduces friction and makes the sands and clays slip easily.

■ The sea washes the slumped material away. Mappleton's cliffs were being eroded at an average rate of 3 m per year. This rapid coastal retreat will continue until the old buried cliff line along the eastern edge of the Yorkshire Wolds is once again exposed. The cliff is composed of much more resistant chalk rock, which will again form impressive white cliffs such as those now existing to the north of Bridlington.

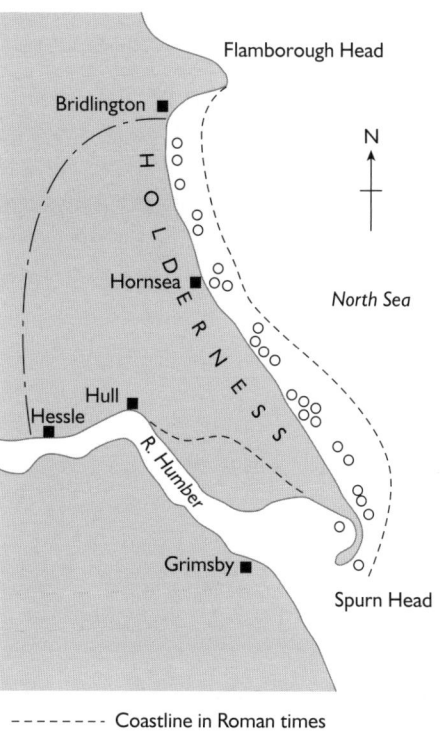

- - - - - - - Coastline in Roman times
○   Lost village or town
— · — Coastline before the Ice Age

0   10   20   30 km

A The lost towns of Holderness.

GREAT BRITAIN GETS SMALLER BY THE DAY

At twilight the burning remains of Sue Earle's clifftop farmhouse at Cowden were reflected in the tide lapping Mappleton Sands below. It had taken 10 hours on Friday to demolish the building, set fire to the broken timbers and clear the site. A pall of smoke drifted over the North Sea, obscuring the flashes from a lighthouse on distant Flamborough Head.

The £250 000 house was destroyed, by court order, because it was on the edge of a clay cliff that had been badly undermined by the sea. It was in danger of sliding onto the beach below. Naturally the law couldn't allow that. So on Friday they pulled the house down and charged its occupants £3500 for the privilege. They now live in a caravan.

This could have been prevented by spending money on coastal defences.

James Warrington, chairman of the parish council, watched smoke rising from the Earles' ruin. The evict-and-demolish policy of East Yorkshire council is seen by him as almost contempt for the land. 'We're losing the very soil and it's probably being washed up on the Dutch coast.' He has researched the fate of his bit of coast. He found that since 1786 the distance between Mappleton church and the cliff edge had been reduced by 3.5 km. In 1990–91 the rot stopped.

'We campaigned strongly and Holderness District Council came up with a scheme for sea defences, using giant rocks – 60 000 tonnes of them – and groynes,' says Mr Warrington. 'By this time a four hectare field of mine had been reduced to two hectares. The scheme has saved Mappleton, but not enough money was spent and our farms are still threatened.'

*From **The Observer** 10 December 1996.*

The New Wider World (Second Edition): Teacher's Resource Book
Neil Punnett and Alison Rae © Nelson Thornes 2003

The Human Cost

Coastal erosion has a human cost. Read the newspaper article, 'Great Britain gets smaller by the day'. What has caused the personal disaster for Sue Earle? She lived at Cowden, 700 m south of Mappleton; she blames the coastal protection measures taken at Mappleton.

The £1.9 million scheme at Mappleton was completed in 1991 (see Figure B). It includes two large rock groynes and a revetment of granite boulders, laid along the beach close to the cliff base. The cliff face was graded and grassed. Although it protects the area of cliff behind the defences, the rate of erosion of the unprotected cliff to the south has trebled. This is because the groynes at Mappleton have destroyed the balance in the coastal system and have stopped the supply of beach material by cutting off the longshore drift. The narrower and much lower beach south of the defences means that waves crash against the cliff foot more often and with more energy. When Figure C has been completed in Question 4a on page 213, the beach profiles clearly show the impact of the groynes.

Hard sea defences such as those at Mappleton do not work in the long term. They have to be regularly repaired and they may cause more damage beyond their limits, as has happened at Cowden. However, such hard defences now protect more than 10% of the British coastline. Further information about the types of coastal protection methods is given in the panel.

Distance from mid-water mark (m)	10	20	30	40	50	60	70	80
Height above mean sea level, north of the southern groyne	1	3	4	4.5	5.5	6	6.5	Cliff foot
Height above mean sea level, south of the southern groyne	1	1.5	2	2.5	Cliff foot			

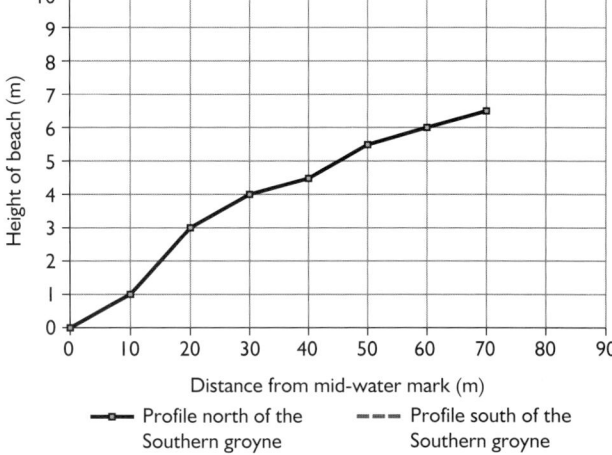

C Beach profiles from Mappleton.

Geographers have tried to persuade local authorities to undertake a policy of managed realignment, whereby the natural processes are allowed to take their course and people affected are paid compensation – this would often be cheaper than the cost of massive hard engineering defence schemes and, in the long term, produce less erosion. The first managed realignment schemes were introduced in the late 1990s.

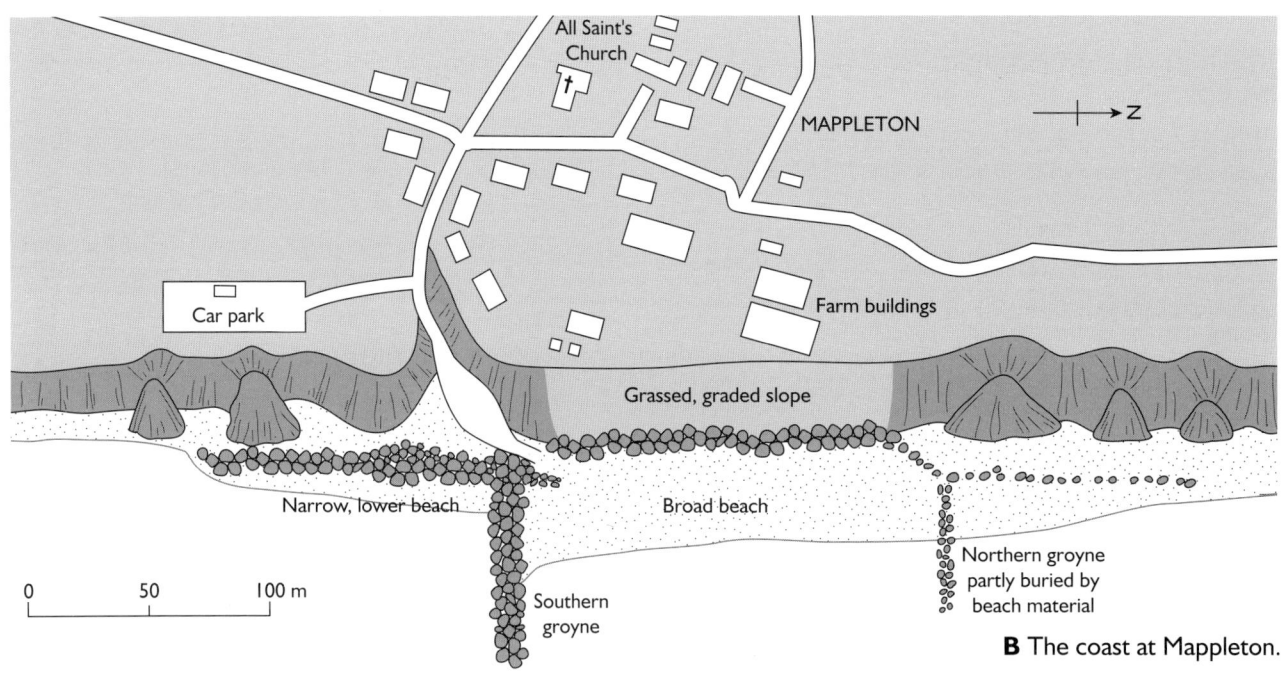

B The coast at Mappleton.

Coastal Protection Methods

Sea wall: made with stone or concrete. It usually has a curved top to reflect wave energy back onto the next advancing wave. A concrete promenade can be built on top of the wall. Usual design life is 50 years, at a cost of £ 5000 per m.

Rock armour: a collection of large interlocking boulders sometimes fixed into position to protect the coast by disrupting the waves. It costs £ 3000 per m to build.

Revetment: there are several types; the traditional revetment is a gently sloping concrete wall that allows the waves to run up it, thereby dissipating their energy. It costs £ 2000 per m to build. A cheaper, but less durable, revetment is a sloping wooden frame (baffle) onto which planks are evenly spaced. It decreases wave energy and allows water and sediment to build up under the planks. The cost to build is £ 500 per m, but does not include repairs. A more effective modern revetment is a permeable design composed of large rocks which can absorb wave energy; this type can, however, be destroyed by storms.

Stone gabions: strong steel cages filled with rocks and some surface sand on which salt tolerant grasses can become established. Although they cost only £ 200 per m, gabions have the disadvantage of being ugly constructions.

Groynes: a long, low wall built out into the sea as near as possible at right angles to the beach. Many of them have to be placed on one beach several hundred metres apart. The aim is to prevent the loss of precious beach sands through longshore drift. Concrete groynes can cost £200 000 each. Wooden groynes are much cheaper, but they rot and are more prone to storm damage.

Beach rebuilding: sand lost through longshore drift is replaced artificially every year. This gives a more natural appearance to the beach, but is expensive – £300 000 per km per year.

Offshore breakwater: a concrete wall (expensive) or interlocking boulders (similar to a rock armour) built a little distance out from the shore protects the coastline by disrupting wave energy and creating an area of calm water inshore. It is, however, rather ugly, it can disrupt the marine ecosystem and can cost £3 million per km.

Do nothing: this might not appear to be a policy for coastal protection, but many people do argue for it on the grounds that nature will take its course in spite of coastal defences. People in coastal homes disagree with this 'policy'.

Managed realignment: recently, and disregarding party politics, most governments and councils, whose responsibility it is to finance any coastal defences, have argued that towns and villages should be protected and farmland should be sacrificed.

Exam Practice Questions

1 a) Where is Mappleton? (1)

 b) What was the rate of erosion per year at Mappleton? (1)

 c) Why is the coastline near Mappleton being eroded so rapidly? (2)

2 How far has the coastline of Holderness retreated since Roman times? (1)

3 What hard engineering works were carried out at Mappleton? (2)

4 Study Figure C again.

 a) Plot the positions for the beach south of the groyne. (4)

 b) Describe and explain the differences in the two beach profiles. (4)

 c) How do the profiles help to explain the different rates of erosion north and south of the groyne? (3)

5 Re-read the newspaper article. Explain the fate of Sue Earle's farmhouse. (5)

Protecting Mappleton

These model answers are designed to show you how to get full marks for the longer questions in the Case Study Extra for this section. Read the question carefully and write your own answer. Then read the model answer and the examiner's notes to see what he or she is looking for when awarding the marks. Decide how many marks you think the examiner would give your answer. Decide how to change your answer to this particular question to increase your marks. What will you do differently next time you answer a similar question?

4 b) Describe and explain the differences in the two beach profiles. (4)

> The beach north of the groyne is higher (up to 6.5 m compared with 2.5 m) and wider (70 m versus 40 m) than the beach to the south of the groyne. The differences are due to the action of longshore drift, which moves material southwards along the coast. The large groyne prevents the beach material moving past it; therefore the beach builds up to the north of the groyne, but is removed to the south.

Level 1 (1–2 marks): *Basic answer, including reference to the beach being higher to the north of the groyne.*
Level 2 (3–4 marks): *Full answer, which includes actual measurements and the effects of longshore drift.*

5 Re-read the newspaper article. Explain the fate of Sue Earle's farmhouse. (5)

> Sue Earle's farmhouse was sited south of the groynes at Mappleton. The groynes, built in 1991, caused beach starvation by preventing the passage of beach material beyond Mappleton. The result was increased erosion at Cowden, undermining the cliff above which Sue Earle's farmhouse stood. In 1996 the farmhouse was demolished by the council because it was in danger of collapsing onto the beach below.

Level 1 (1–2 marks): *Basic answer, including reference to erosion of the cliff.*
Level 2 (3–5 marks): *Full answer includes beach starvation, caused by the groynes at Mappleton, and the demolition of the house by the local council.*

CHAPTER 19 is divided into five sections: the glacial system, processes and landscapes; glacial landforms; transportation and deposition by ice; glacial landforms on OS maps; and glacial highlands and human activity.

The Glacier System 📖 316

The concept of the glacier as a system centres on the inputs, stores, flows and outputs involved. The terms *accumulation* and *ablation* are introduced.

Activity Sheet 19.1 provides a diagram of the glacier system, accompanied by a graph completion exercise showing the relationship between accumulation and ablation through a year.

Erosion 📖 316

Frost shattering creates the moraine with which a glacier erodes its valley. The two main processes, abrasion and plucking, are described.

Glacial Landforms 📖 317 – 318

Corries, arêtes, pyramidal peaks and glacial troughs are studied in detail, with diagrams and photographs to support the descriptions.

Activity Sheet 19.2 includes the features of a corrie and a contour map of a glaciated area from which pupils have to identify glacial landforms.

Activity Sheet 19.3 tests pupils' understanding of more glacial landforms through the use of diagrams.

Transportation and Deposition 📖 319

Types of moraine are described, followed by erratics and drumlins. Figure 19.11 is an excellent photograph of a terminal moraine and it can be easily linked to the diagram, Figure 19.10.

Activity Sheet 19.4 links directly with page 319 in the pupil book. The main activity involves interpreting two cross sections in order to check pupils' understanding of the difference between fluvial and glacial deposits.

Glacial Landforms on OS Maps 📖 320

The two OS maps on this page, one from Snowdonia, the second from the Lake District, are used to show examples of glacial landforms. Grid references are given. A useful exercise would be to ask pupils to draw sketches of the landforms as they are shown on the OS maps.

Case Study 19 Glacial Highlands and Human Activity 📖 321

Numerous examples are used to show the attractions of glacial highlands for human activities and the effects of these activities.

Activity Sheet 19.5 includes a map of glacial deposits in Denmark, and goes on to question pupils' understanding of human activities in glaciated highland areas.

Case Study Extra A Glacier in the Canadian Rockies

The Athabasca Glacier is probably the most photographed glacier in the world because of its proximity to the Icefields Parkway highway. In summer, more than 10 000 people pass this spot every day. The case study explains the glacier's formation and its dramatic retreat since the 1840s. The exploitation of the glacier for tourism is discussed in detail. The glacier's uncertain future is a good point to end the *Teacher's Resource Book* – in Geography there are more questions than answers.

1 Study Figure A. Place the following labels in the correct boxes on the sketch. (3)

> • Meltwater   • Precipitation   • Evaporation   • Avalanches   • Zone of ablation   • Zone of accumulation

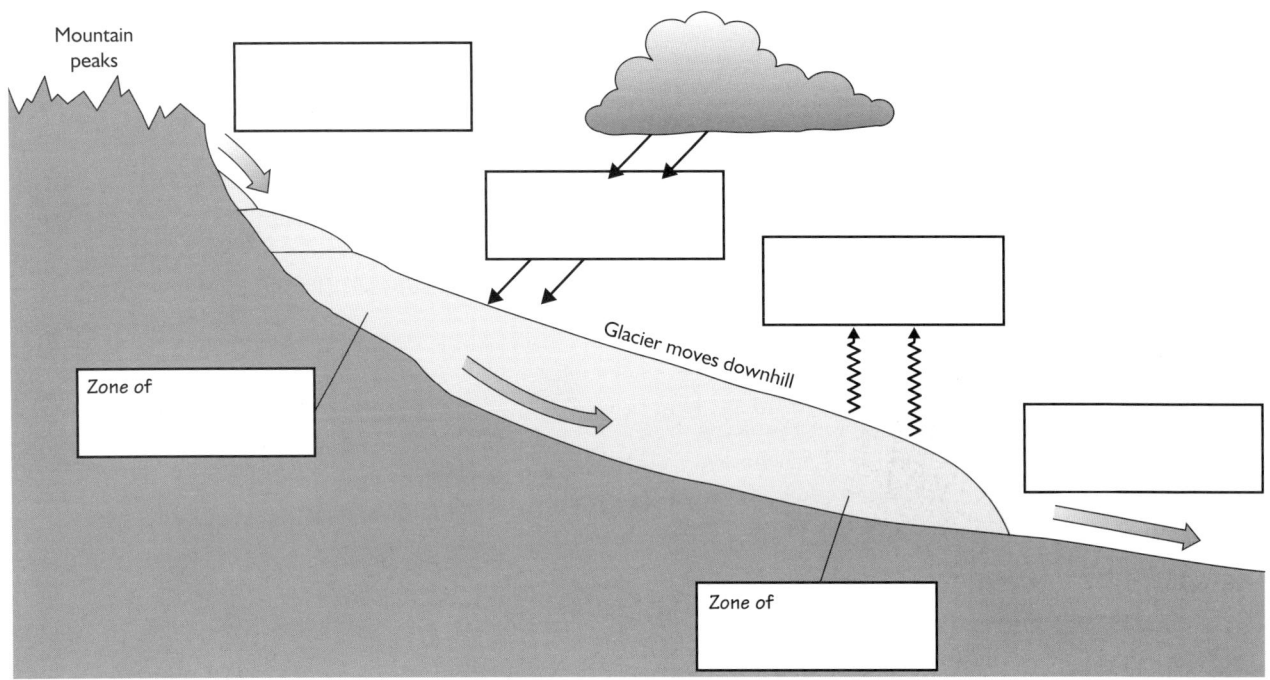

A The glacial system.

2 How is glacier ice formed? (3)

3 Study Figure B.

  a) What do you understand by (i) accumulation and (ii) ablation? (2)

  b) Complete the graph by using the following data. (5)

Month	August	September	October	November	December
Accumulation	3.5	5	7.5	12	16
Ablation	13.5	12	7.0	3	2

  c) Describe and explain the pattern of accumulation and ablation shown on the graph. (4)

  d) (i) Calculate the total accumulation and ablation for the year. (2)

  (ii) Is this glacier advancing or retreating? (1)

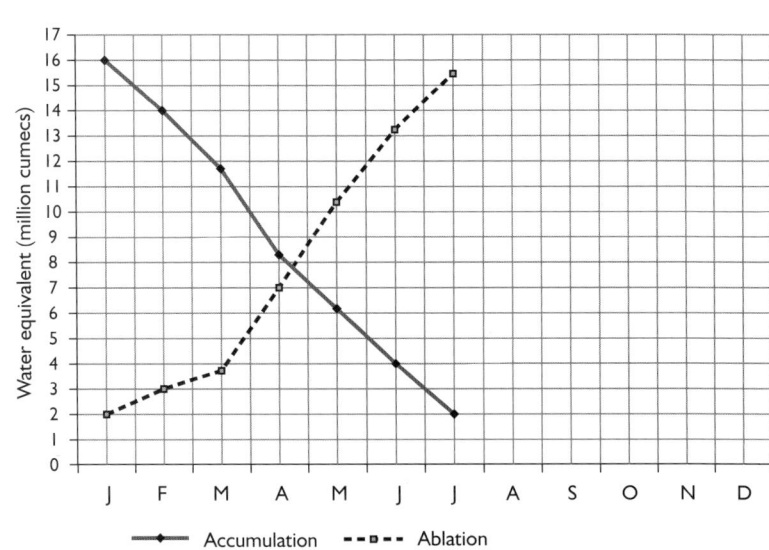

B The glacier budget for a Norwegian glacier.

The New Wider World (Second Edition): Teacher's Resource Book
Neil Punnett and Alison Rae © Nelson Thornes 2003

1 Study Figure A.

a) Name the landform shown. *(1)*

b) Place the following labels in the correct boxes on the diagram. *(5)*

> • Abrasion   • Ice-fall   • Plucking   • Rock lip   • Freeze-thaw

c) Describe the processes of glacial abrasion and glacial plucking, using page 316 in the pupil book to help you. *(4)*

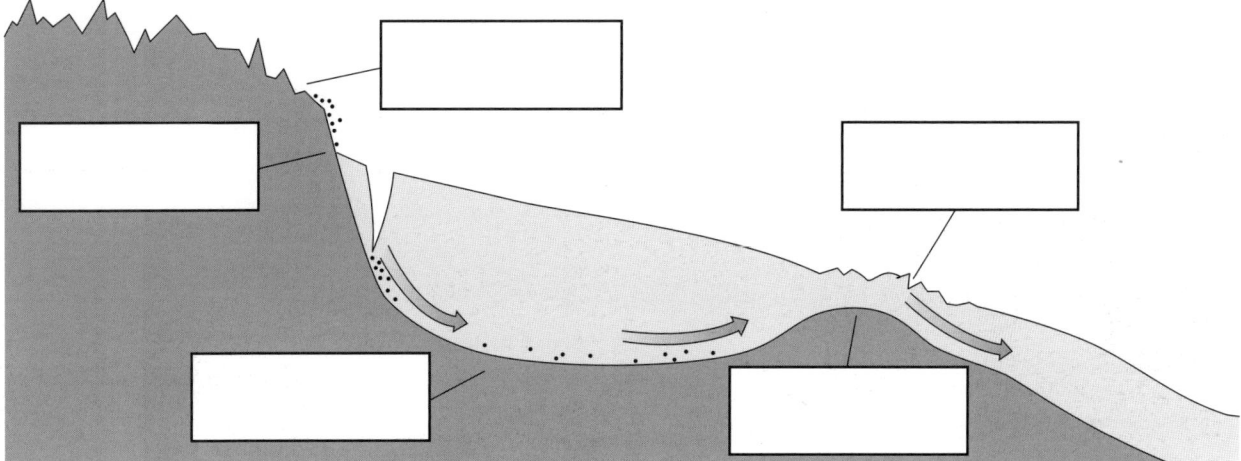

A A corrie.

2 Figure B is a contour map of a glaciated area in the UK.

a) Name two glaciated highland areas in the UK. *(2)*

b) Identify the five glacial landforms labelled A to E on the map, choosing from the following list. *(5)*

- Tarn
- Arête
- Hanging valley
- Ribbon lake
- Pyramidal peak
- Corrie truncated spur

c) Choose one of the five landforms A to E and, with the aid of a labelled diagram, describe how it may have been formed. *(5)*

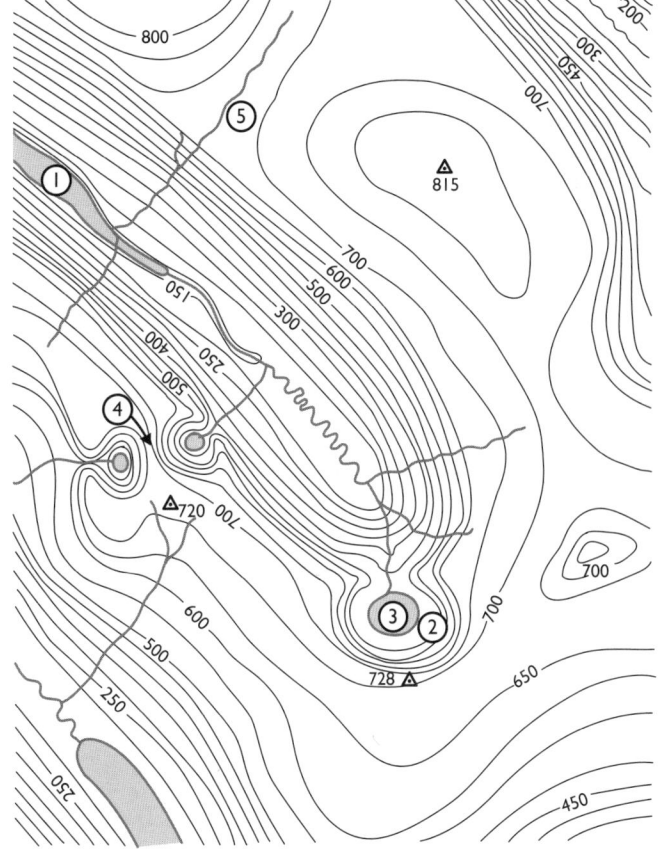

B A glaciated highland area.

1 a) Study Figure A. Name the landforms at A, B and C. *(3)*

b) Explain how landforms A and B were formed. *(4)*

2 Describe and explain the different appearance of the valley in Figure B after glaciation. *(6)*

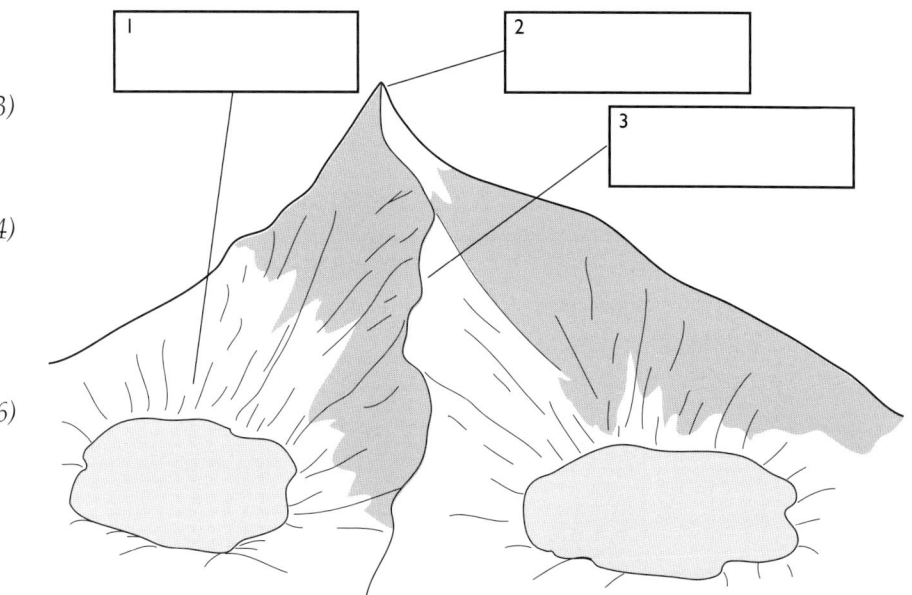

A A pyramidal peak, arête and corrie.

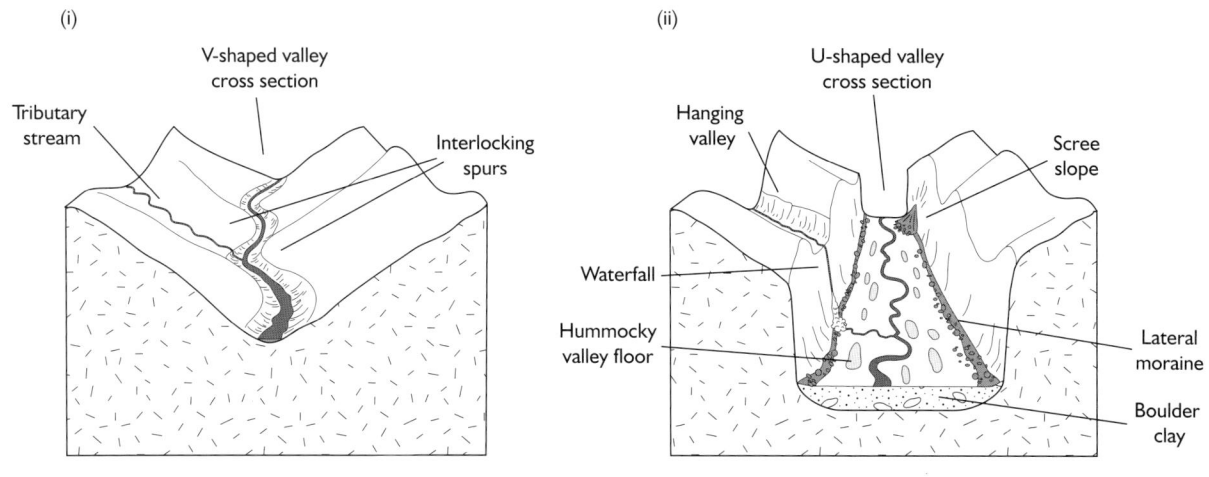

B A valley (i) before and (ii) after the Ice Age.

3 a) What is moraine composed of? *(2)*

b) Identify the five types of moraine shown as D to H on Figure C and explain their locations. *(5)*

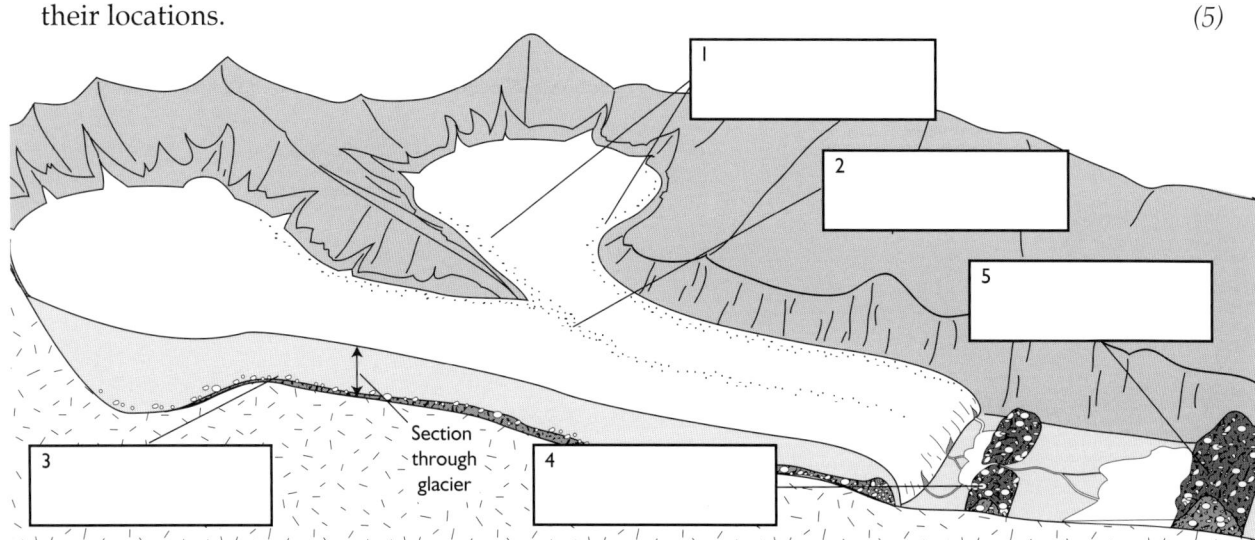

C Types of moraine.

1 Draw a labelled sketch diagram to show how glacial moraine can lead to the formation
   of a ribbon lake in a glacial trough. (3)

2 a) What is a drumlin? Illustrate your answer with a diagram. (3)

   b) How are drumlins believed to have been formed? (3)

3 a) What is an erratic? (1)

   b) How does the study of erratics help in understanding the direction of ice movement? (2)

4 Study Figure A, which shows sections through the deposits in two valley floors.

   a) Which of the deposits do you think has been made by ice and which by water? (2)

   b) Give three reasons to explain your choice. (6)

(i)

(ii)

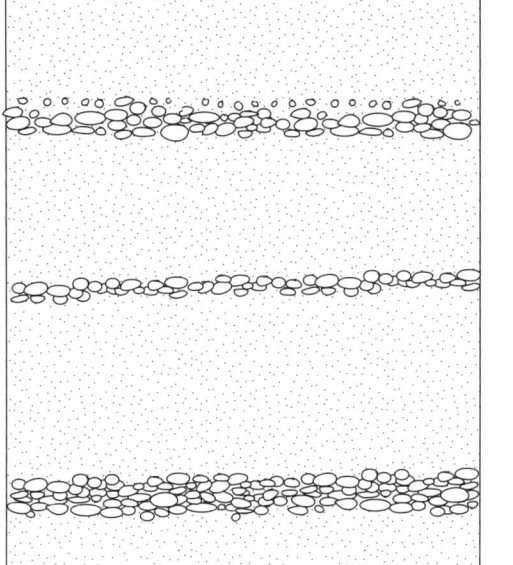
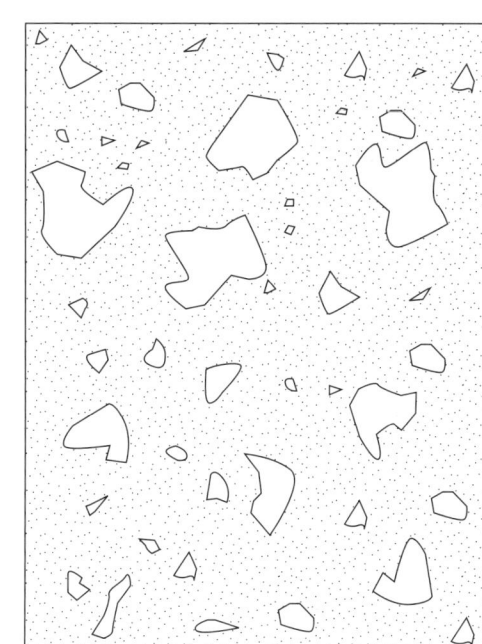

A Vertical sections through the deposits in two valley floors.

1 Study Figure A, which shows features of glacial deposition in Denmark.

   a) Using evidence from the map, state the direction of ice movement during the glacial period. Explain your answer.   *(3)*

   b) Describe the appearance, drainage and fertility of the soils likely to be found at A, B and C.   *(6)*

A The glacial deposits in Denmark.

2 For each of the following, describe ways in which the landscape of a glaciated highland area causes problems and suggest how these problems might be overcome:
(i) communications; (ii) settlement; (iii) farming.   *(9)*

3 a) Name one glaciated highland area you have studied.   *(1)*

   b) Describe and explain the attractions of the area for tourists.   *(4)*

   c) List three problems caused by tourism for the local inhabitants.   *(3)*

A Glacier in the Canadian Rockies

The Athabasca Glacier is in Jasper National Park in the Rocky Mountains of Alberta in Canada. It is an outlet glacier from the Columbia Icefield, the largest area of ice and snow in the Rockies. The icefield covers an area of nearly 350 sq km, a similar size to the Isle of Wight. Surrounded by 11 of the Rockies' highest peaks and located at an average elevation of 3000 m, it is fed by about 10 m of snowfall annually. In places the ice is 900 m thick. The Athabasca Glacier itself covers 6 sq km. Its depth varies from 90 to 350 m.

The Athabasca Glacier flows through a dramatic U-shaped glacial trough. The glacier has retreated and advanced since the height of the Ice Age 15 000 years ago. In about 1300 A.D., the glacier had shrunk to be only about 3 km long. The global climate was cooling, however, and the glacier started to re-advance. It surged down the valley, overwhelming a forest of spruce and fir that had grown up during the earlier period of relative warmth. The effects of this most recent advance are clearly evident in the high lateral moraines along the valley walls. By 1840, the Athabasca Glacier stretched far down the Sunwapta River Valley. In the latter half of the 19th century, the climate started to become warmer again and the ice slowly began to melt, leaving a large terminal moraine to mark its farthest extent. Since then, it has receded by about 1500 m. The rate of retreat varies from year to year. When the retreat of the glacier's snout halts for a while, recessional moraines are formed. There are several recessional moraines running across the valley between the current position of the snout and the upper car park. The retreating glacier has exposed extensive areas of smoothed and striated bedrock.

The ice melts faster than it can be replaced and the snout of the glacier retreats. But because it is on a mountain slope, the ice in the glacier still flows downhill. It is flowing at 125 m per year at the icefall, but only 25 m per year at the snout. It is important to understand that, because the rate of ablation has generally exceeded the rate of accumulation in the recent past, as a whole the glacier has shrunk substantially and the position of the snout is retreating up the valley. The recessional moraines serve as markers of this retreat. The glacier's lateral moraines now stand 150 m or more above the ice surface, showing how much melting has occurred in recent decades.

Figure A shows that the Athabasca Glacier actually joined with the neighbouring Dome Glacier at the maximum extent of its advance in the mid-19th century. The earliest photograph of the glacier dates from 1906 and was found by Brian Luckman and a team of geographers from the University of Western Ontario. Figure B on page 222 shows sketches based on that photograph and another taken from the same spot in 2002. Note the appearance of the lake, which began forming in 1940; Sunwapta Lake is a pro-glacial meltwater lake. In addition to the retreat of the snout, the glacier appears to be over 100 m less thick in 2002 than it was in 1906.

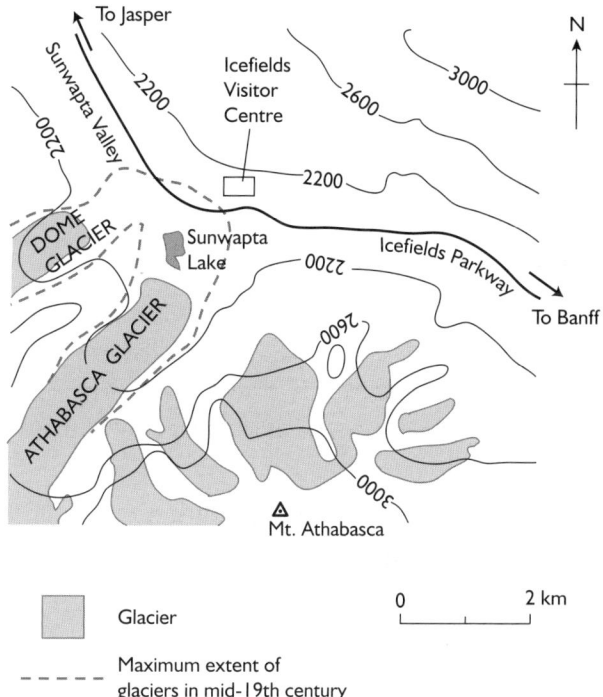

A The Athabasca Glacier.

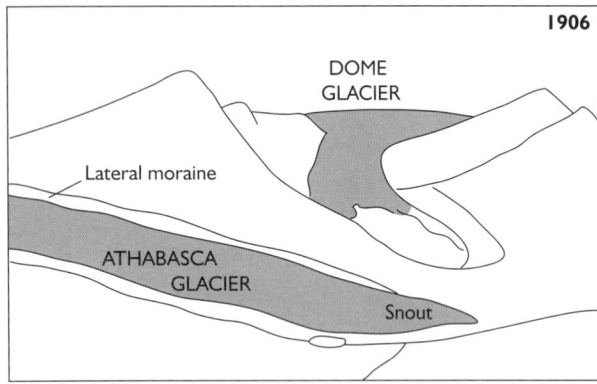

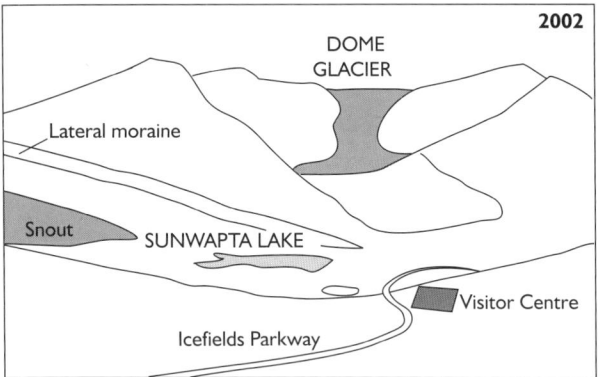

B Sketches of the Athabasca Glacier based on the first known photograph taken in 1906 and another taken from the same spot in 2002.

Tourism

The Athabasca Glacier is probably the most photographed glacier in the world because its snout lies close to the busy Icefields Parkway road (Highway 93) linking the towns of Banff and Jasper. More than 10 000 people pass this spot every day in the summer. Most will stop at the Columbia Icefield Centre, an impressive £3 million complex opened in 1996, which replaced a smaller lodge dating from 1931. The Icefield Centre has extensive car parking for 730 cars and 100 buses. Inside the centre, models and audio-visual displays explain the formation and movement of the glacier. The Centre also includes a restaurant, café and hotel rooms for overnight stays. On busy days, more than 6000 people will take a trip on snocoaches – specially designed six-wheel drive vehicles with huge balloon tyres, which take them out onto the glacier where they can disembark and walk on the glacier's surface. Experienced guides also offer hikes across the glacier and up the icefall onto the Icefield above. Helicopters fly wealthier tourists directly up to the Icefield.

Concern has been expressed at the effects on the area of millions of trampling feet, especially on the newly exposed land at the glacier's snout. The Icefield Centre, its extensive car parks, the snocoach road and base, the noisy helicopters and road vehicles are all obtrusive in this frozen landscape. However, the benefits of concentrating the tourists at one point probably outweigh the disadvantages.

The Athabasca Glacier has been designated a World Heritage Site by the United Nations and it attracts over one million visitors each year.

The Future

Research suggests that the Athabasca Glacier is shrinking by 30% every 100 years and could be gone in less than 300 years. Other glaciers in the area could disappear in less than 50 years and scientists warn that the melt could speed up if the earth's temperatures continue to climb. If the glaciers in the Columbia Icefield disappear, so will a critical water supply for western Canada. The Athabasca Glacier feeds several large prairie water systems. In a hot dry summer like the summer of 1998, the glacier meltwater is the only thing that keeps the rivers flowing.

Exam Practice Questions

1 Figure C is a sketch of the Athabasca Glacier.

　a) Name the features shown by the letters A to D, choosing from the following list.　*(4)*

- lateral moraine
- terminal moraine
- outwash sands and gravels
- boulder clay deposits.

　b) Choose two of the features and describe how each one was deposited.　*(4)*

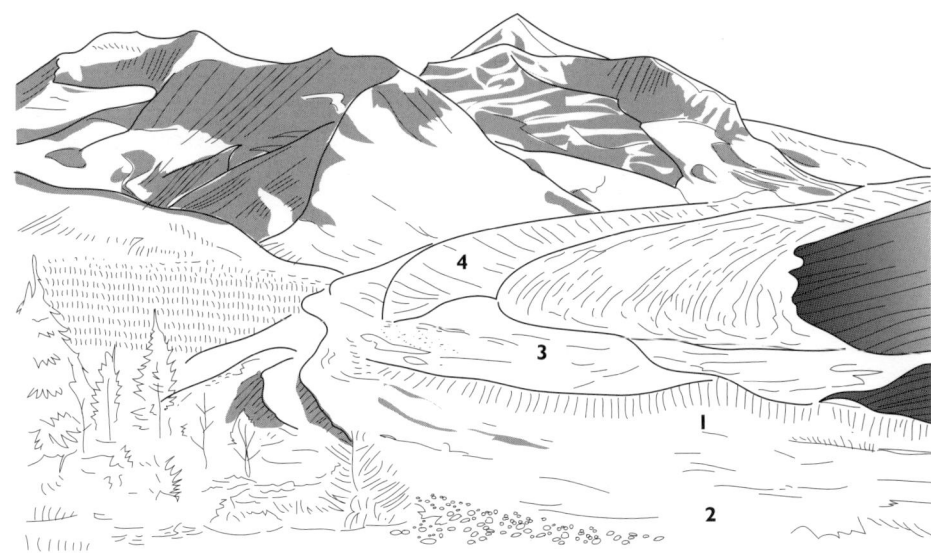

C Features of the Athabasca Glacier.

2 Figure D shows the retreat of the snout of the Athabasca Glacier since 1850.

　a) Complete the graph by plotting the statistics in the table below for 1950, 1975 and 2000.　*(3)*

Year	1850	1875	1900	1925	1950	1975	2000
Distance from maximum extent (m)	0	110	260	470	770	1320	1610

　b) Describe the trends revealed by the graph.　*(3)*

　c) Study Figures A and B. What evidence is there of the glacier's retreat?　*(3)*

3 a) How has the Athabasca Glacier's potential for tourism been exploited?　*(6)*

　b) What are the conflicts between tourism and the natural environment in the Athabasca Glacier area?　*(4)*

D Distance of the Athabasca Glacier from its maximum extent, since 1850.

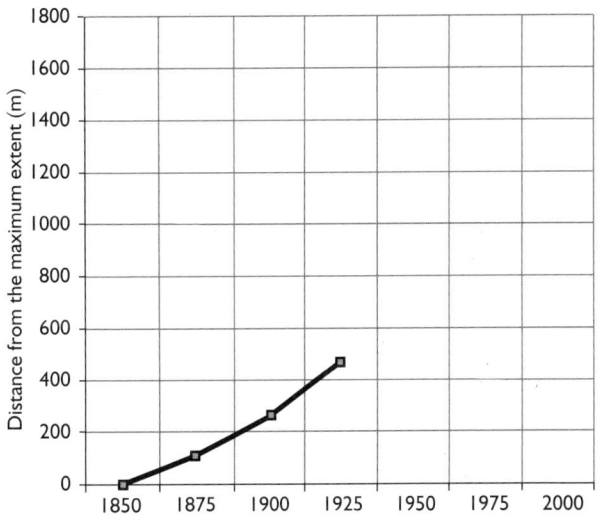

A Glacier in the Canadian Rockies

These model answers are designed to show you how to get full marks for the longer questions in the Case Study Extra for this section. Read the question carefully and write your own answer. Then read the model answer and the examiner's notes to see what he or she is looking for when awarding the marks. Decide how many marks you think the examiner would give your answer. Decide how to change your answer to this particular question to increase your marks. What will you do differently next time you answer a similar question?

3 a) How has the Athabasca Glacier's potential for tourism been exploited? *(6)*

A large information centre called the Columbia Icefield Centre has been built on the highway near the snout of the Athabasca Glacier. The centre includes car parking, dining and accommodation, in addition to multimedia displays explaining the formation and movement of the glacier. People can walk from the car parks up to the snout of the glacier and can climb up onto the ice. Up to 6000 people per day travel on specially designed snocoaches, which drive onto the Athabasca Glacier. Passengers can get out and walk on the ice. Tourists can hire guides to lead them across the glacier and up to the Icefield above. Wealthier tourists can take helicopter flights directly to the Icefield.

> **Level 1 (1–2 marks)**: *Basic answer including reference to tourists walking on the ice.*
> **Level 2 (3–4 marks)**: *More detailed answer including the Columbia Icefield Centre and the snocoach trips.*
> **Level 3 (5–6 marks)**: *Full, detailed answer including details of the Icefield Centre's facilities.*

b) What are the conflicts between tourism and the natural environment in the Athabasca Glacier area? *(4)*

The Athabasca Glacier is one of the most visited glaciers in the world – over one million visitors each year. The tourists bring jobs and useful income into this region. However, large-scale tourism does affect the fragile glacial environment. The tourists walk across the newly exposed land at the glacier's snout and create broad, muddy trampled areas. The Icefield Centre and its large car parks are out of place in this wilderness area. The noise from the cars, snocoaches and helicopters break the natural silence. On balance, the benefits of concentrating the tourists at one, easily accessible point, probably outweigh the disadvantages; the rest of the Columbia Icefield and its glaciers receives far fewer visitors as a result.

> **Level 1 (1–2 marks)**: *Basic answer, including reference to the large number of tourists and cars.*
> **Level 2 (3–4 marks)**: *More detailed answer, including the impact on the environment of the Columbia Icefield Centre, the snocoach and the helicopter trips, set against the income and jobs created by tourism.*